# R Statistics

AF625957

NIPA® GENX ELECTRONIC RESOURCES & SOLUTIONS P. LTD.
New Delhi-110 034

www.nipaers.com
Browse, Search, Read & Buy...
Print Books eChapters
eBooks Publishers
Forthcoming Subject Catalogues
New Titles
Online Resources on:
✓ Current Affairs
✓ Reasoning, Logic & Aptitude
✓ Competitive Examinations
✓ English Language Lab
✓ Writing & Pronunciation Tools
✓ Personality Development
✓ Online Programmes for Professionals
✓ Interview Preparation
Pay Using
PhonePe Google Pay Paytm PayPal UPI CCAvenue

# R Statistics

**D.K. Samual**
Principal Scientist
Indian Institute of Horticultural Research
Hessaraghatta Lake Post
Bangalore, Karnataka

NIPA® GENX ELECTRONIC RESOURCES & SOLUTIONS P. LTD.
New Delhi-110 034

NIPA® GENX ELECTRONIC
RESOURCES & SOLUTIONS P. LTD.

101,103, Vikas Surya Plaza, CU Block
L.S.C. Market, Pitam Pura, New Delhi-110 034
Ph : +91 11 27341616, 27341717, 27341718
E-mail: newindiapublishingagency@gmail.com
web: www.nipabooks.com

***For customer assistance, please contact***
Phone: + 91-11-27 34 17 17
Fax: + 91-11- 27 34 16 16
E-Mail: feedbacks@nipabooks.com

© 2023, Publisher

ISBN: 978-81-19254-76-7

All rights reserved. No part of this publication may be reproduced, stored in a retrieval system or transmitted in any form or by any means, including electronic, mechanical, photocopying recording or otherwise without the prior written permission of the publisher or the copyright holder.

This book contains information obtained from authentic and highly reliable sources. Reasonable efforts have been made to publish reliable data and information, but the author/s, editor/s and publisher cannot assume responsibility for the validity, accuracy or completeness of all materials or information published herein or the consequences of their use. The work is published with the understanding that the publisher and author/s are not attempting to render any professional services. The author/s, editor/s and publisher have attempted to trace and acknowledge the copyright holders of all material reproduced in this publication and apologize to copyright holders if permission and/or acknowledgements to publish in this form have not been taken. If any copyrighted material has not been acknowledged, please write to us and let us know so that we may rectify the error, in subsequent reprints.

**Trademark Notice:** NIPA®, the NIPA® logos and their presentations (the way they are written/ presented) in this book are the trademarks of the publisher and hence may not be used without written permission, if copied or used without authorization, the infringer will be prosecuted as per law.

NIPA® also publishes books in a variety of electronic formats. Some content that appears in print may not be available in electronic books, and vice versa.

Composed and Designed by NIPA®.

# Preface

My aim in writing this short book is to enable an average student like me to graph and chart the data in an understandable and analytically correct manner using R. I have a wild dream, that high school students to Post-Doctoral researchers of all fields will use R and that this book will help them learn the first steps, so I decided not to write extra and not to repeat what was already available on the internet/print. You may feel that the book is fast paced, Yes it is, it is meant to be read sitting before a PC and to enter the commands. Practice makes perfect and the best way to learn R is to work with it and ask questions when you don't get the results. Very special thanks to the staff of NIPA for agreeing to publish this book in color. Hope this book on R will help you to start on a journey of discovery with R. Good Luck and God Bless, I hope that soon I will be able to see charts based on this book get published in research journals. One request please work on the code and try till you succeed. Don't give up easily. For comments/suggestions/criticisms, please contact on email: dksamuel2@gmail.com.

Author

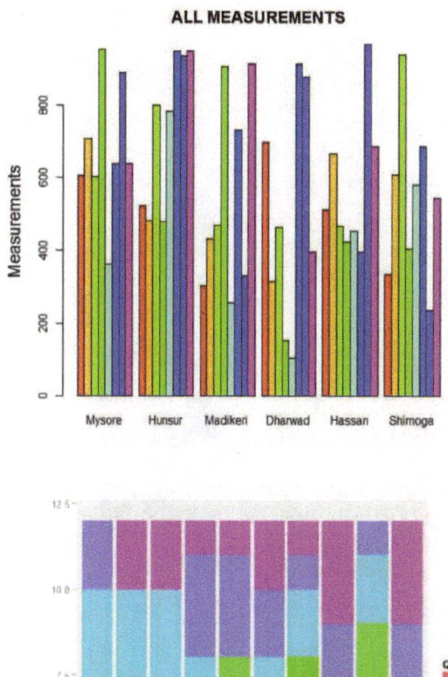


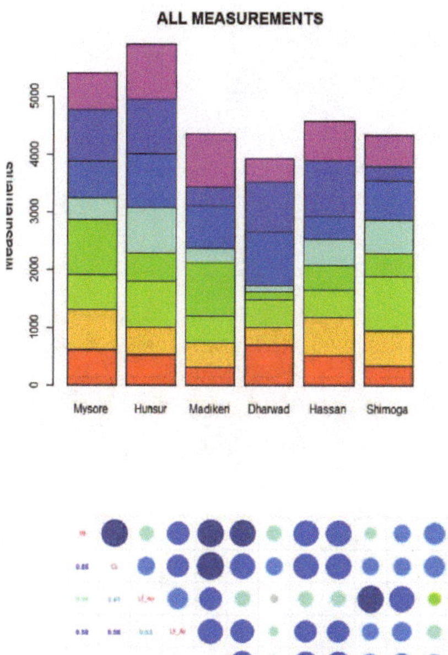


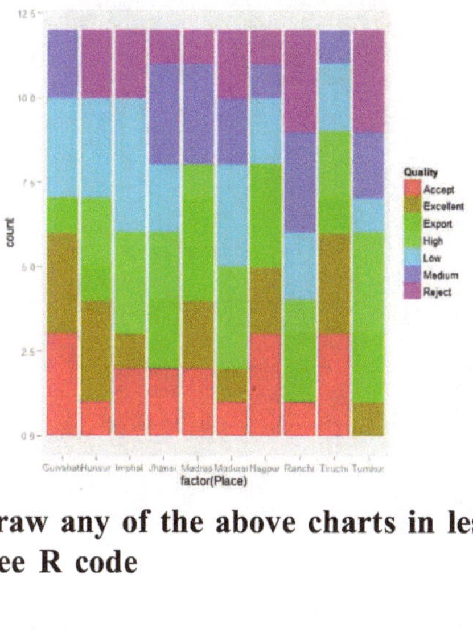


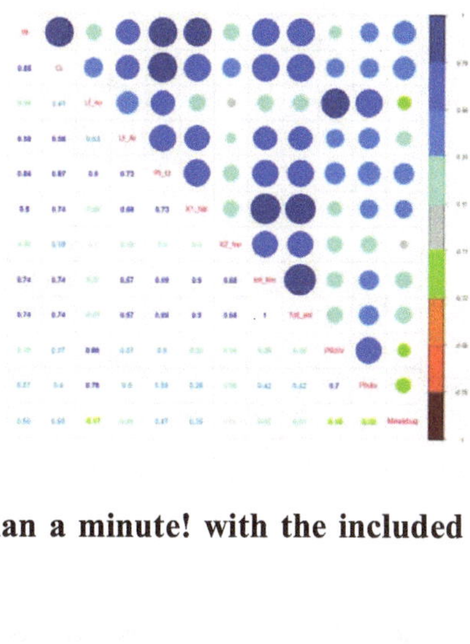

**Draw any of the above charts in less than a minute! with the included free R code**

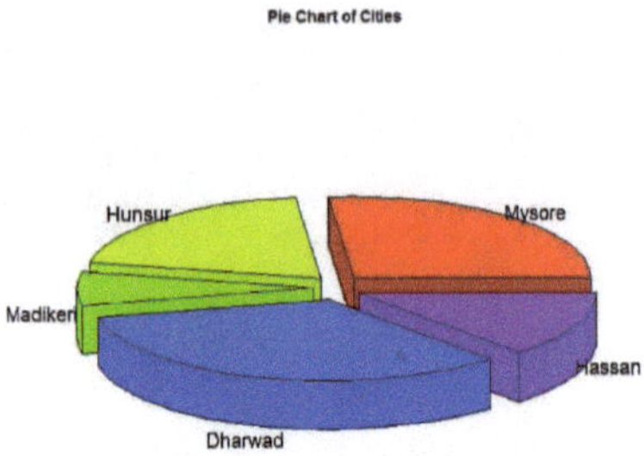


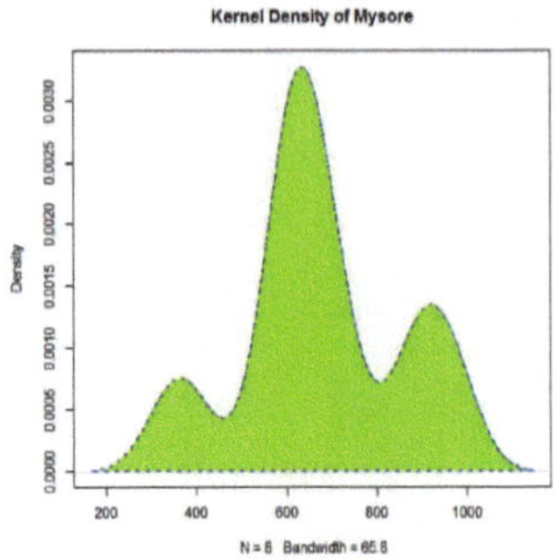


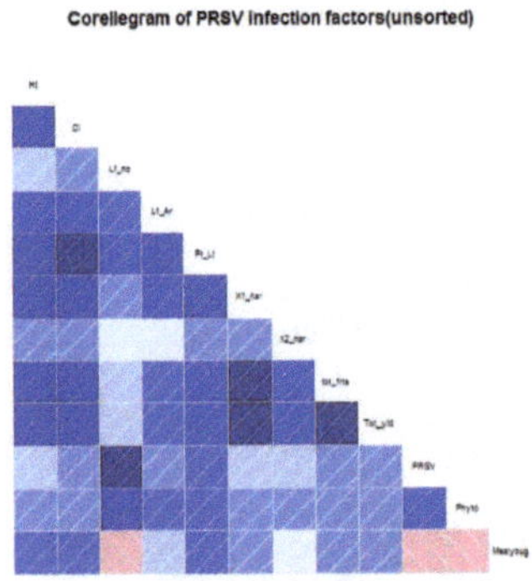


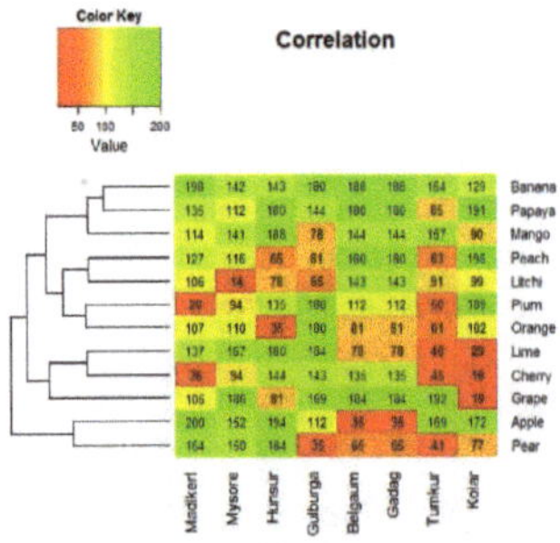


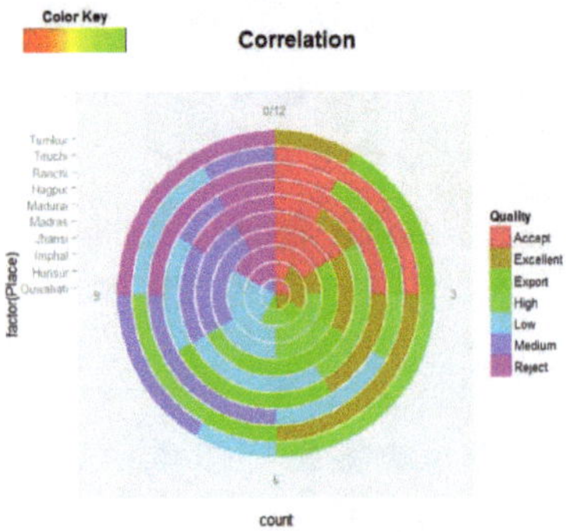


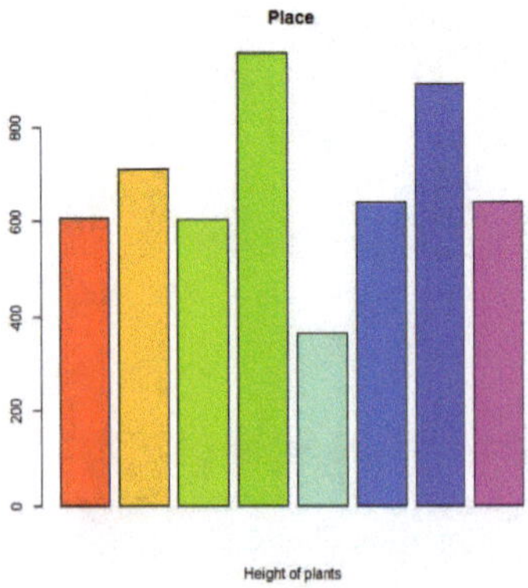


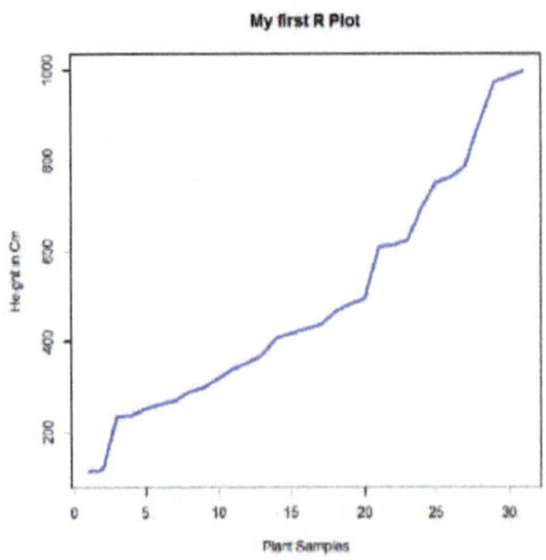


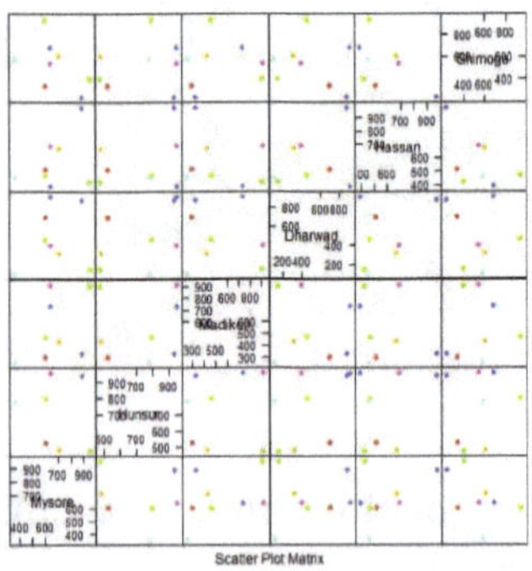


**Draw any of the above charts in less than a minute! with the included free R code**

R is a software package / language / environment for statistical analysis and graphics.

**It Reads Any Type of Data** *.txt, .csv, .dat, etc.), You can also scrape (collect) data from websites and execute SQL queries **It Supports Large Data Sets. Reproducibility.** R allows you to add comments to your scripts to make it clear to you and others on what you're doing as data and analysis are separated in R. As R is fully **free you can use the latest version on any / as many computers so you and your student can work on the same version.** You can use R on Windows, Mac, Linux, and Unix. **Anyone (Including You) Can Contribute Packages to the Community to Improve its Functionality.**

You can download R from : http://cran.r-project.org. R has a command line interface and that is a huge hurdle for many users, but believe us you will find that using a command line is also easy. However we strongly recommend the R Studio text editor for working with R scripts (http://www.rstudio.com/ide/download).

**Note:** R and few packages, with R Studio is available in the supplied DVD

***R Charting in less than 1-minute Ok let us start.***

Type or even better cut and paste the following text lines in a R > prompt.

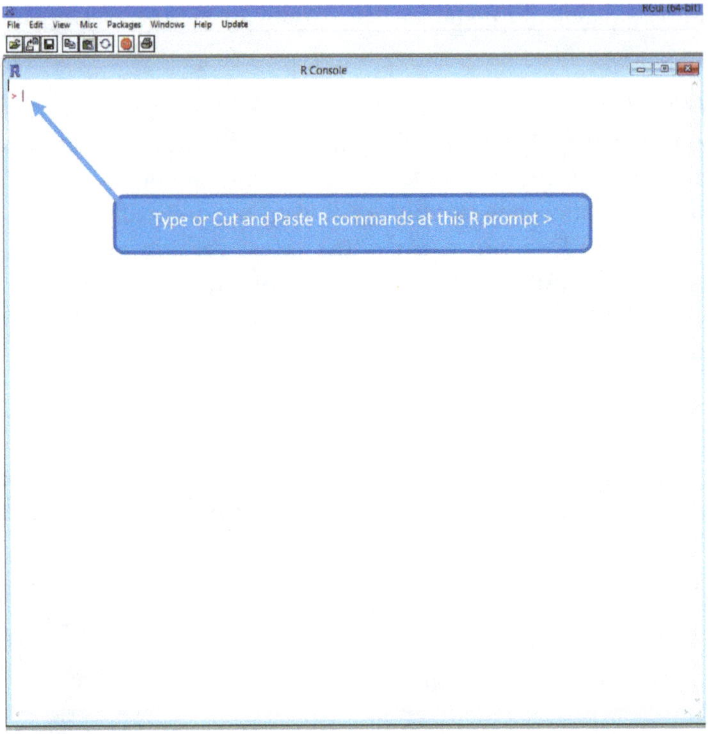


*Please do not include the symbol of R prompt* > so Type or cut and paste from x <-...... and plot ...... Please do not put any spaces / commas, as R is extremely specific about its syntax.

**Note:** The Installation of R is easy and has been extensively covered in the Appendix. We assume and are sure that you have installed R successful in your computer and are sitting facing a R terminal.

! Don't include this > symbol when you type the code. Please type / cut and paste carefully

**>x<-c(113,117,235,252,263,271,290,300,321, 340,999)**

**# Now press the <ENTER KEY>, After you press the <ENTER # KEY> the R prompt > will reappear**

**>plot(x,type="l",col="blue",lty=1,lwd=3, xlab="Plant Samples", ylab="Height in Cm", main="My first R Plot")**

# After the R prompt reappears then type / cut and paste the second line and press the <ENTER KEY> again.

You can type / cut and paste both the lines together and press the <ENTER KEY> once

! Don't include this > symbol when you type the code. Please type / cut and paste carefully. Type like this

**> x <-c(113,117,235,252,263,271,290,300,321,340,999) press the <ENTER KEY>**

**After the R prompt reappears**

**> plot(x, type="l",col="blue",lty=1,lwd=3, xlab="Plant Samples",ylab= "Height in Cm", main="My first R Plot") press the <ENTER KEY> again**

**# you can also do like this**

**> x <-c(113,117,235,252,263,271,290,300,321,340,999)**

**plot(x, type= "l",col="blue",lty=1,lwd=3, xlab="Plant Samples", ylab="Height in Cm", main="My first R Plot") press the <ENTER KEY>**

**Now the chart will appear**

See your first chart in the graphics window which will appear adjacent to the code window on

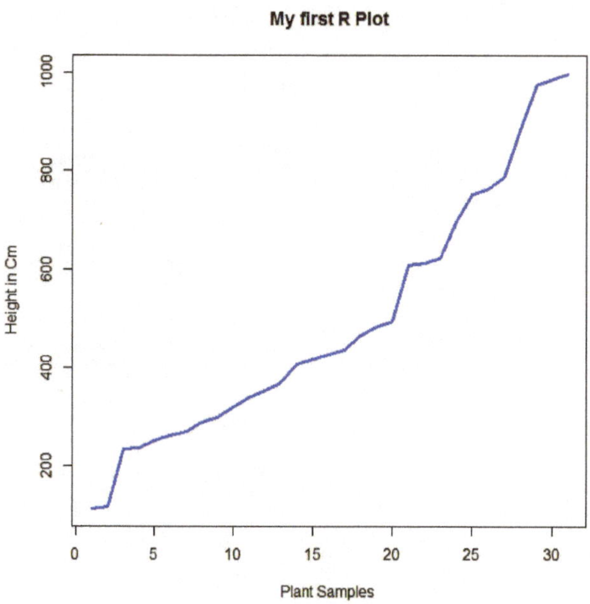

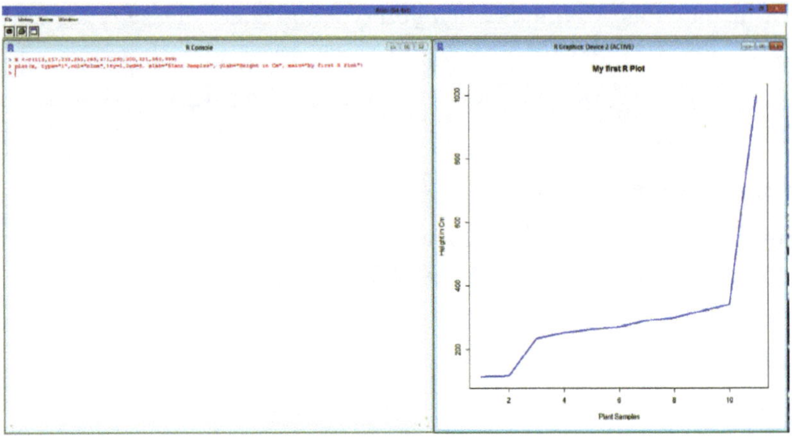


Adjacent to the code window the chart will appear on the R-graphics window, **Note:** The color of the tilte bar has changed to blue indicating that it is active (alive)

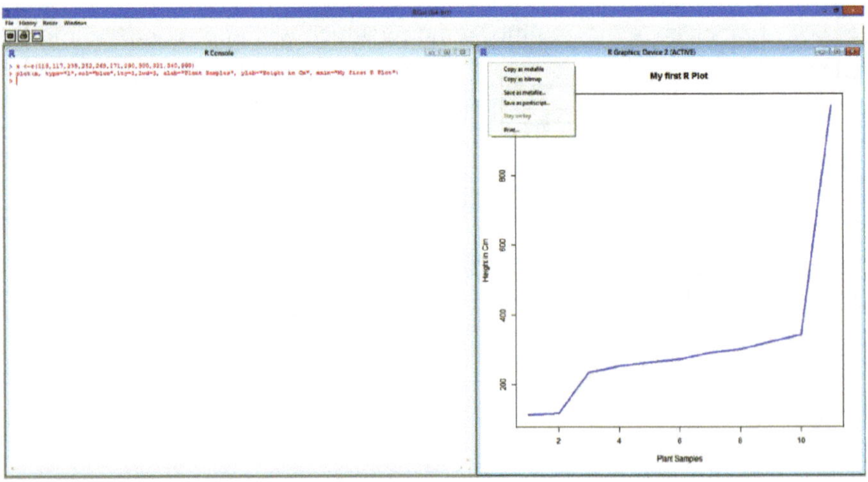

*Code Explained:*

# As the values have been pre-sorted from smallest to largest in a spread sheet the curve will be http://www.researchrockstar.com/wp-content/uploads/2011/03/bigstock_Graph_analysis_11233205.jpg smooth, without any jumps / spikes.

#

**You have 2 options, 1. To save the chart, 2. To copy the chart**

**1. To save the chart, click on File in the Menu at the R code window and select Save as to save in 7 different file formats.**

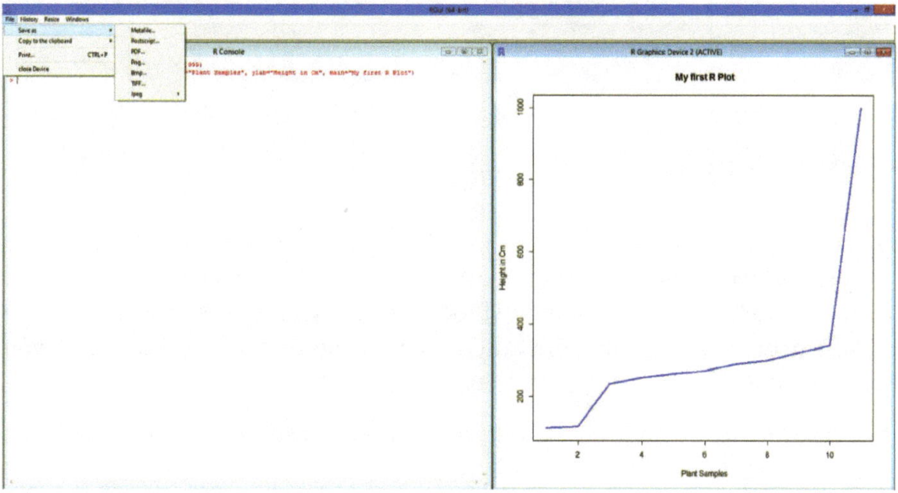


To copy the chart Place the cursor on the top margin of the chart (below the title), Right click on the mouse, from the menu, select copy as metafile and paste in MSWord

**If the data points have been unsorted, the chart will have spikes**

**x <-c(290,113,300,999,271,252,263,117,235,321,340)**

**plot(x, type="l",col="blue",lty=1,lwd=3, xlab= "Plant Samples", ylab="Height in Cm", main="My first R Plot - unsorted")**

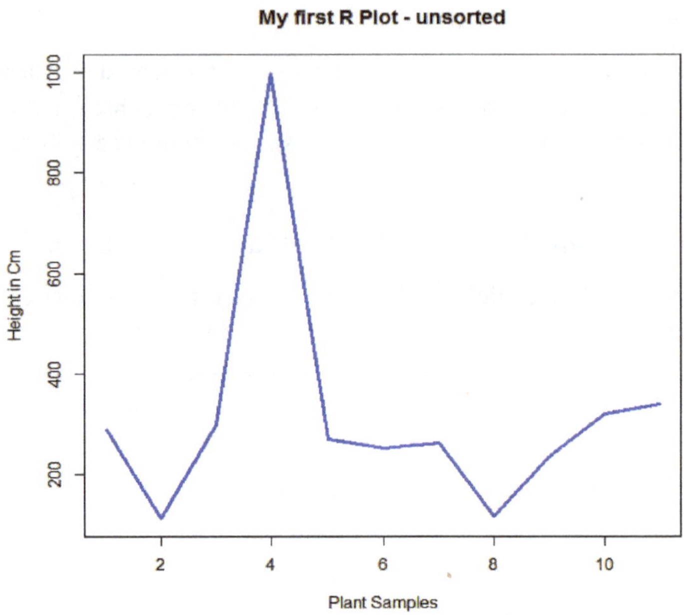


Before you use R intensively, it is most important that the data is made available in a form which is easily understood both by you and R. For data entry spreadsheets are most suitable, after entering the data in a spreadsheet you can save it as a coma separated files (*.csv)

For ease of use, we suggest the following free open software when using R to prepare the data prior to analysis in R

Use the following free office suites (with equivalents to MSWord, MS Excel and MS Powerpoint). OpenOffice(http://www.openoffice.org),

Libre office ( http://www.libreoffice.org)

Kingsoft (http://www.kingsoftstore.com/software/kingsoft-office-freeware).

You can also use Gnumeric (which is a spreadsheet only) http://www.gnumeric.org

The produced *.csv data files can be easily edited or cleaned up in

Notepad plus plus [Notepad++] (http://notepad-plus-plus.org)

Atom (https://atom.io)

Pspad (http://www.pspad.com)

**Note:** All the above software is available in the supplied DVD

*The required data files are available as csv files in the folder researchrdata present in the DVD. Copy the entire folder to your C drive as*

C:/researchrdata (if you want to copy to any other location, please see that there are no spaces in between the file names) and change the code accordingly.

*How to Get Data from a spreadsheet into R*

Although there are R packages for importing Excel data directly we advise you to export the spreadsheet to a .csv file and then import the .csv file in R

# Easier method to read a csv file

Use the R file chooser command: When you use the file choose command in R, it will open up an Explorer style file manager, and you can navigate to your csv file and select it

**> my.data = read.table(file.choose(), header=TRUE)**

*Code Explained:* The data in a csv file used in R will be arranged in columns, with the first row having the header. This is informed to R by the command header=TRUE

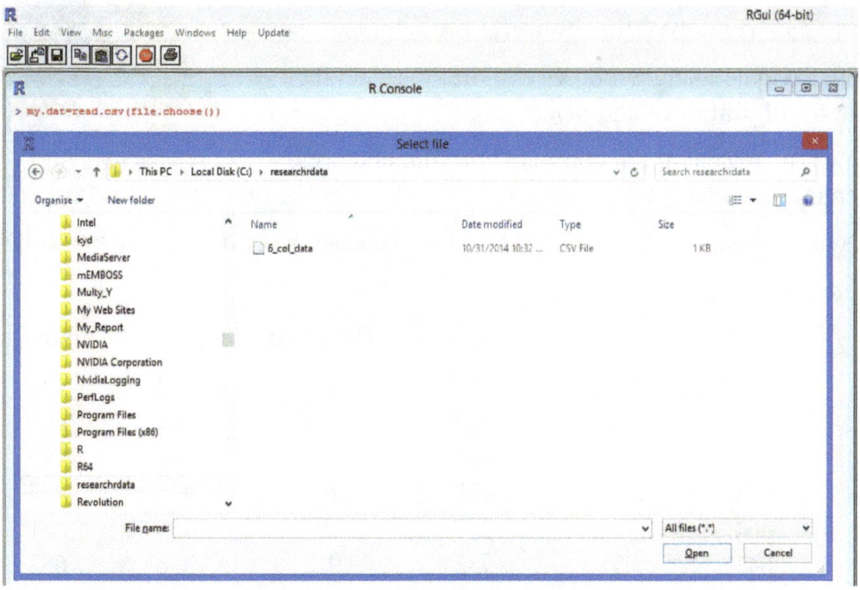


# Second Method to read a csv file using the read.csv() function.

code for importing a .csv file in R and storing it in an object called 'mydata':

**> mydata <- read.csv("C:/researchrdata/ 6_col_data.csv")**

**# use the # liberally to record your comments, R will ignore text following a #**

**# Please be careful of the direction the / face, it must be forward in MSWindows,**

**# Please remember to use the inverted commas "" properly, You can use either single '' or double "" quotes but please stick to a uniform style and don't mix both types.**

***Code Explained:*** You have made an object called as mydata and have told R that from now onwards mydata will represent the read.csv file. The text within the "" informs R where the file read.csv is present on the computer.

> **edit(my.data)**

***Code Explained:*** This command will open up a spreadsheet with the csv file arranged in it; you can do rudimentary editing in it

***DATA REQUIRED***

***Second R Chart in less than 1-minute***

DATA REQUIRED

Cut and paste the following 3 lines at the R prompt >

**>my.dat=read.csv(file.choose())** press<enter> key

now navigate using windows explorer to select the file **6_col_data.csv** and select it

now type my.dat at the command prompt and press <enter> key and

R will respond by displaying the **6_col_data.csv with its proper headings.**

> my.dat

| | Mysore | Hunsur | Madikeri | Dharwad | Hassan | Shimoga |
|---|---|---|---|---|---|---|
| 1 | 607 | 524 | 304 | 699 | 513 | 336 |
| 2 | 709 | 481 | 432 | 315 | 667 | 609 |
| 3 | 603 | 801 | 471 | 464 | 468 | 940 |
| 4 | 955 | 479 | 908 | 151 | 425 | 404 |
| 5 | 364 | 784 | 257 | 105 | 455 | 581 |
| 6 | 641 | 951 | 733 | 915 | 396 | 688 |
| 7 | 891 | 936 | 332 | 880 | 968 | 236 |
| 8 | 641 | 951 | 915 | 396 | 688 | 545 |

Type the attach command so the data can be assessed by their column names

> attach(my.dat)

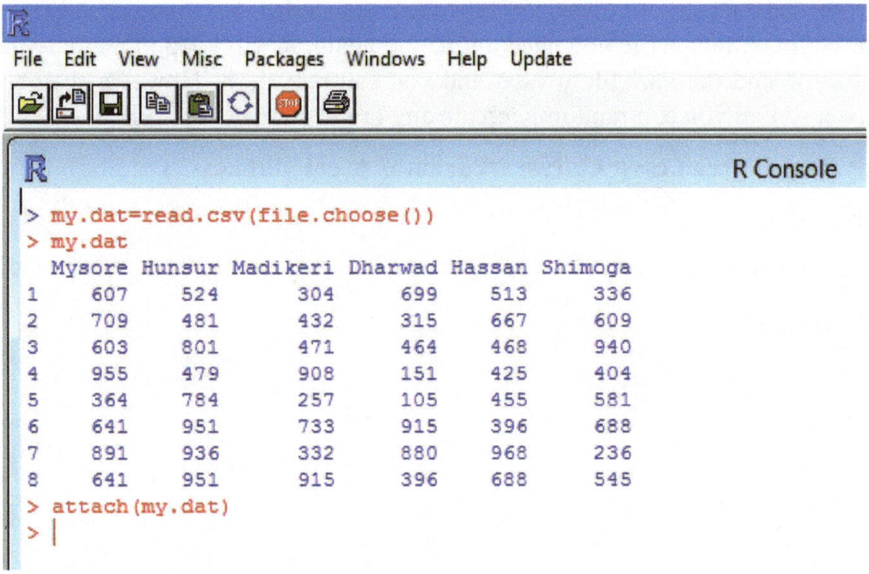


Now type the command

> **boxplot(my.dat,col=rainbow(6))**

You will get the graph in the graphic window

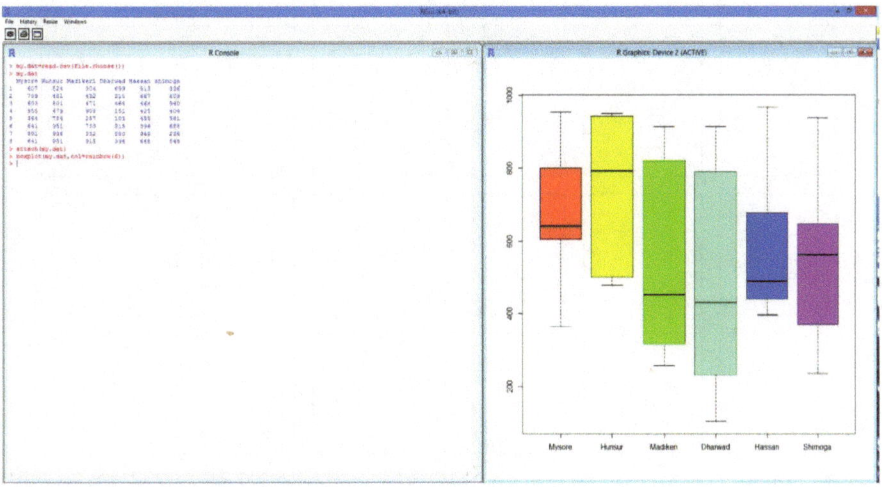


Save the graph / copy, it as earlier discussed

To see the data structure visually type the command, as you have already read the 6_col_data.csv file you need not read it again, R will keep in memory the data file and the variable my.dat and you can recycle it. However they are given so that you can randomly create any graph at any time.

**my.data <- read.csv("C:/researchrdata/ 6_col_data.csv")**

**my.dat**

**attch(my.dat)**

**plot(my.dat)**

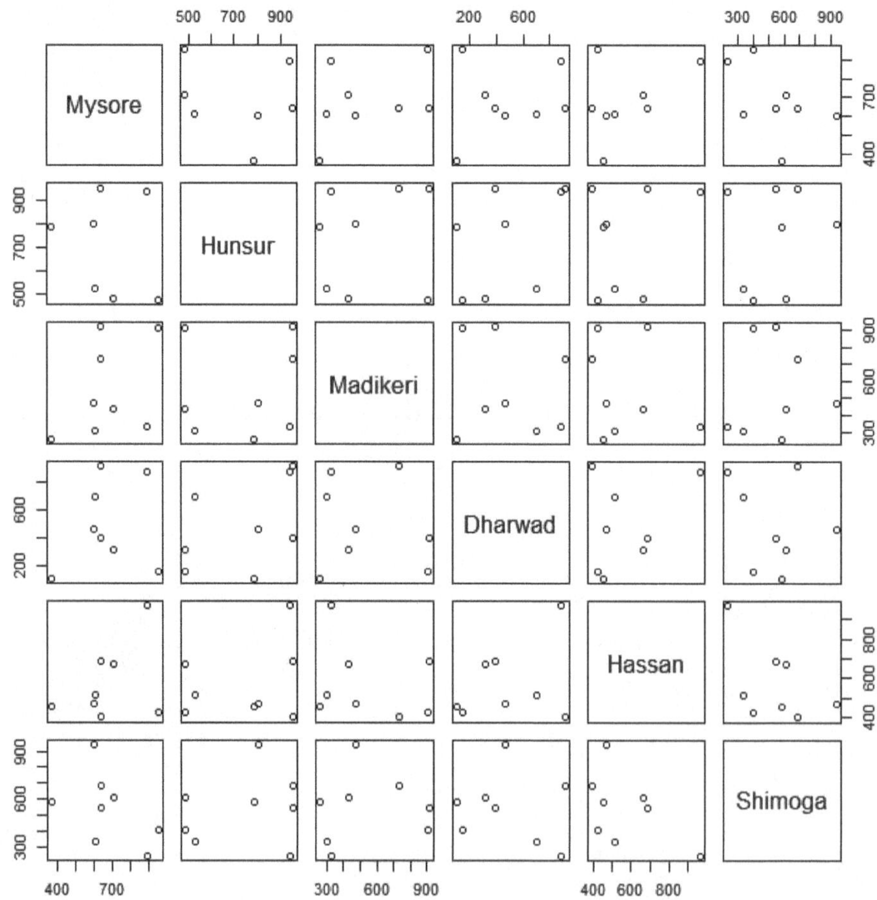

The data indicators are not in colour so you can add colour to the graph by this command

**my.data <- read.csv("C:/researchrdata/ 6_col_data.csv")**

**my.dat**

**attch(my.dat)**

**plot(my.dat,col=rainbow(8))**

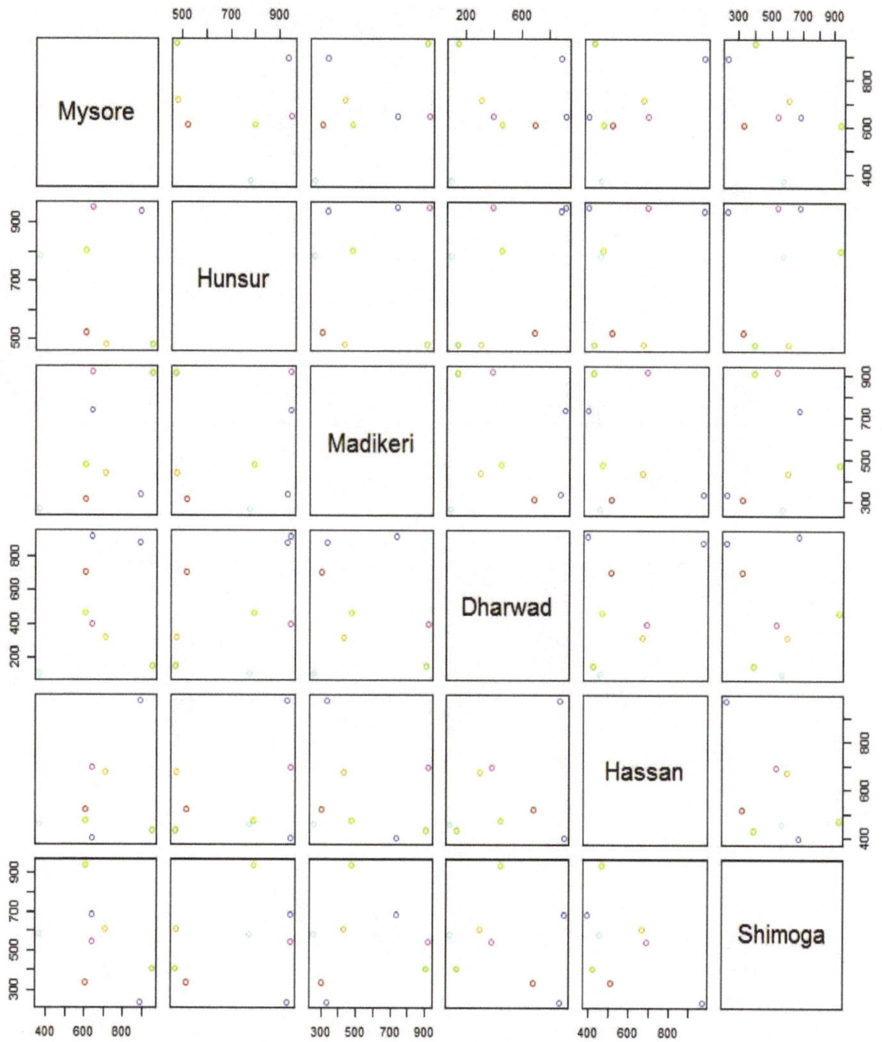

To add solid markers issue this command

```
my.data <- read.csv("C:/researchrdata/ 6_col_data.csv")
my.dat
attch(my.dat)
plot(my.dat, pch = 16,col=rainbow(8))
```

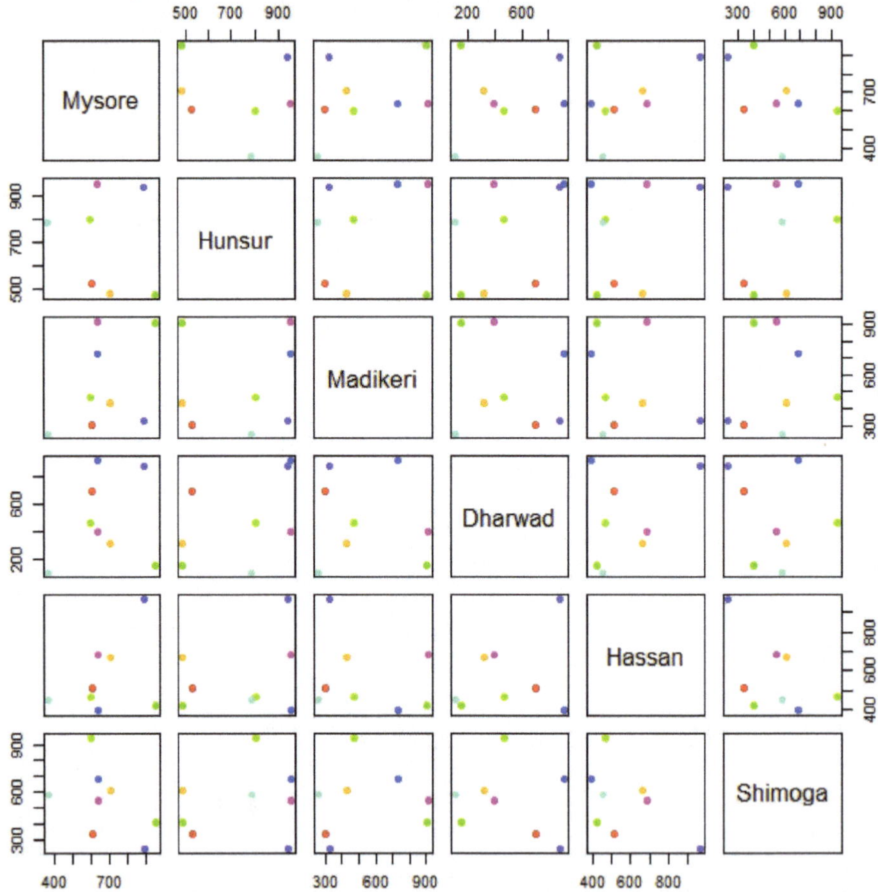

Issue the following commands as variations

**my.data <- read.csv("C:/researchrdata/ 6_col_data.csv")**

**my.dat**

**attch(my.dat)**

**plot(my.dat, pch = 16, col=heat.colors(8))**

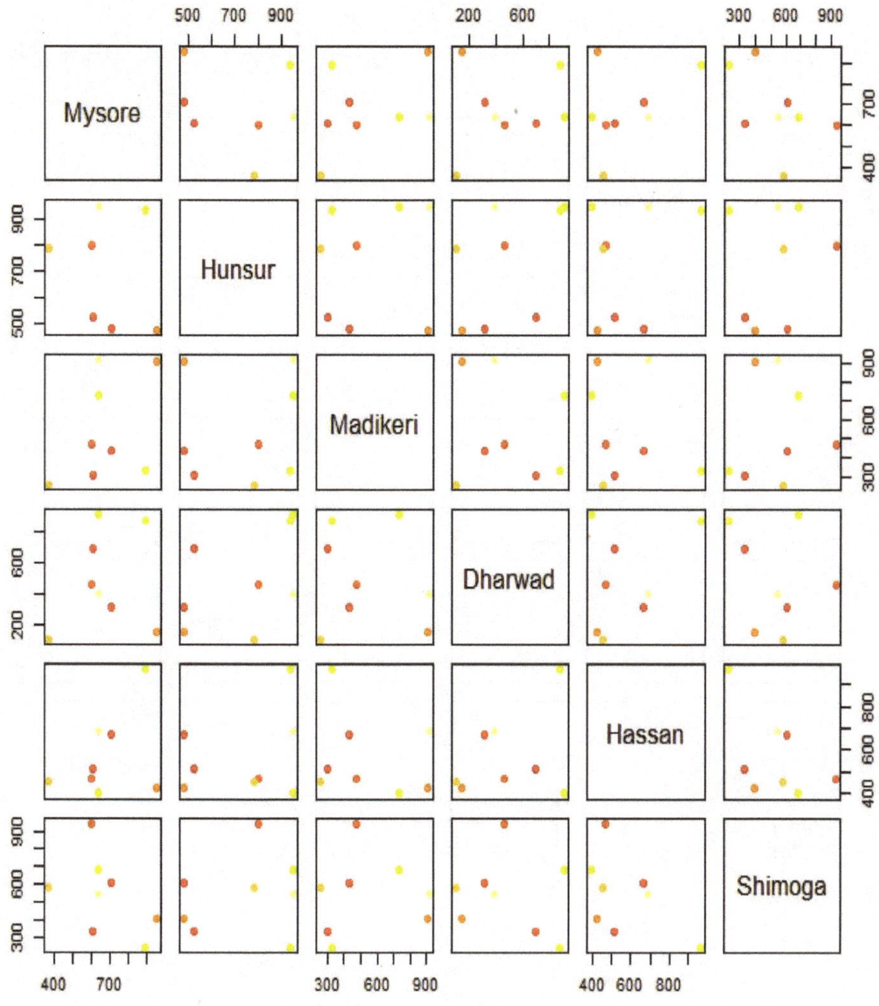

```
my.data <- read.csv("C:/researchrdata/ 6_col_data.csv")
my.dat
attch(my.dat)
plot(my.dat, pch = 16, col=terrain.colors(8))
```

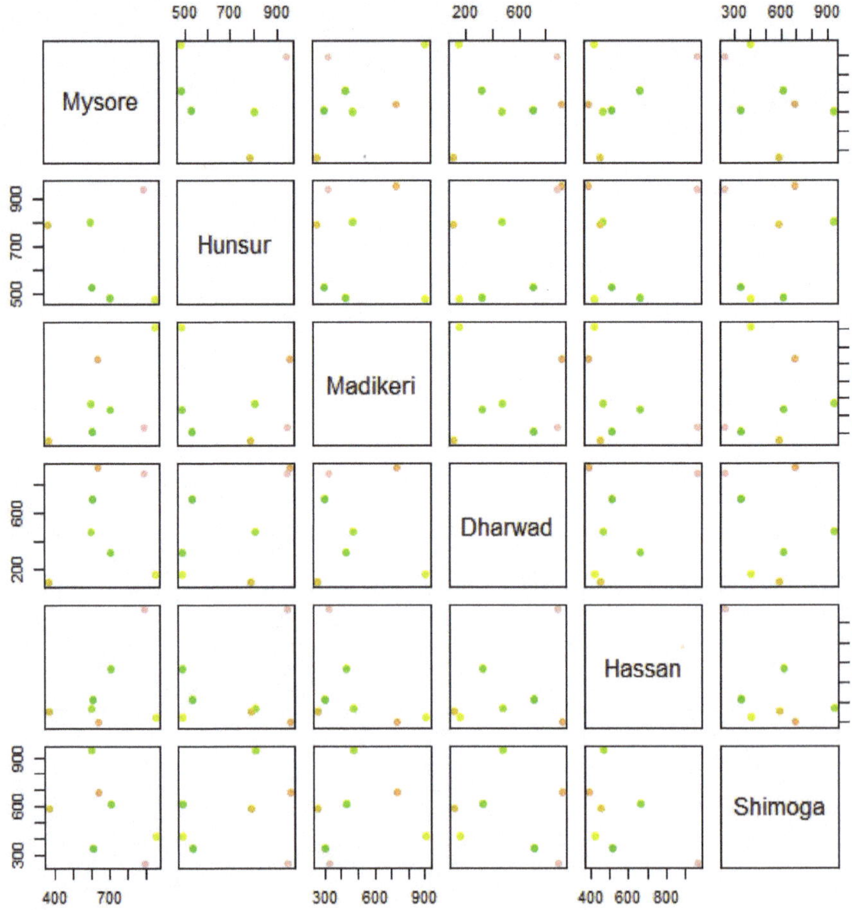

```
my.data <- read.csv("C:/researchrdata/ 6_col_data.csv")

my.dat

attch(my.dat)

plot(my.dat, pch = 16, col= topo.colors (8))
```

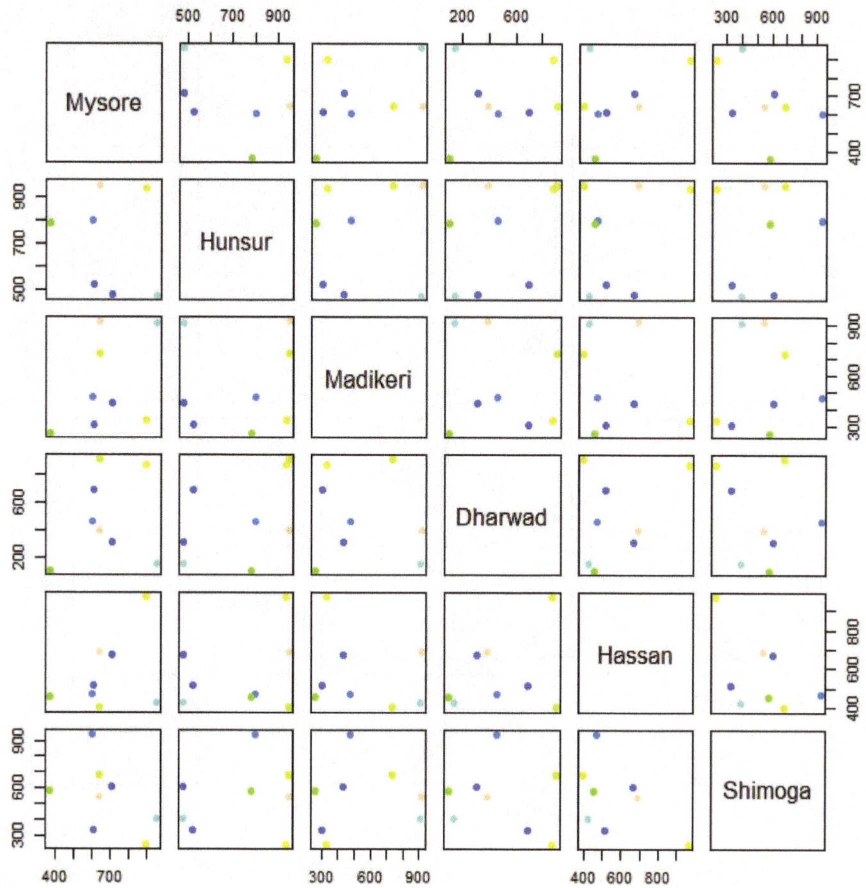

**my.data <- read.csv("C:/researchrdata/ 6_col_data.csv")**

**my.dat**

**attch(my.dat)**

**plot(my.dat, pch = 16, col= cm.colors (8))**

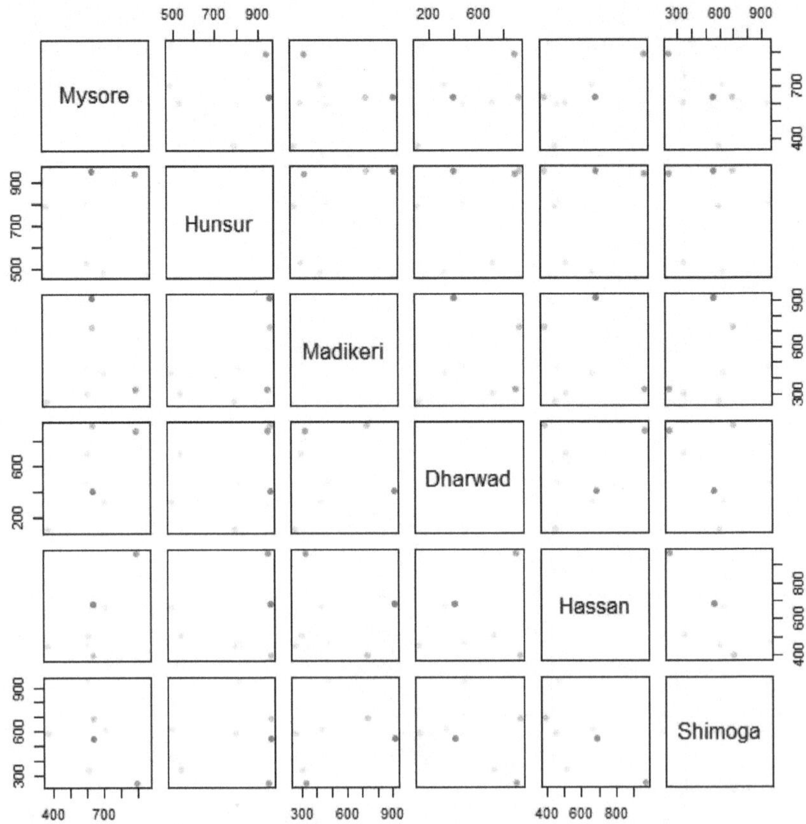


By now you would have seen that the program R is a well behaved friendly program controlled by commands issued at the command line, As you progress you will find that this command is helpful and you can scroll through a list of commands given in a R session by pressing the <up> or <down> arrows.

Let us embellish the plot more by adding title (captions). Label the axes, Remember embellishments are dependent on the type of the chart.

**my.data <- read.csv("C:/researchrdata/ 6_col_data.csv")**

**my.dat**

**attch(my.dat)**

**barplot(as.matrix(my.dat), main="ALL MEASUREMENTS", ylab = "Measurements", cex.lab = 1.5, cex.main = 1.4, beside=TRUE, col=topo.colors(8))**

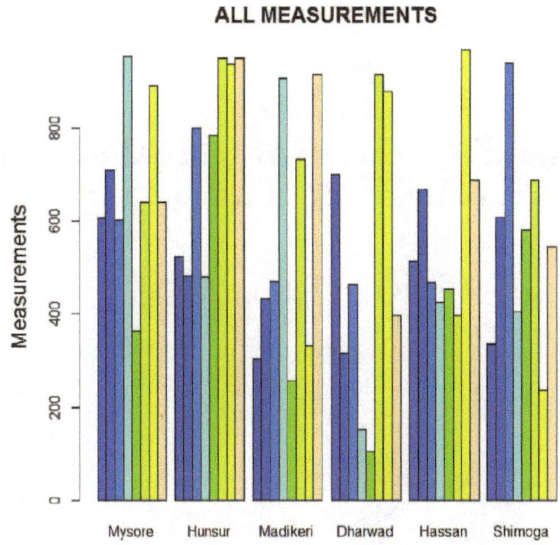


**my.data <- read.csv("C:/researchrdata/ 6_col_data.csv")**

**my.dat**

**attch(my.dat)**

**barplot(as.matrix(my.dat), main="ALL MEASUREMENTS", ylab = "Measurements", cex.lab = 1.5, cex.main = 1.4, beside=TRUE, col=rainbow(8))**

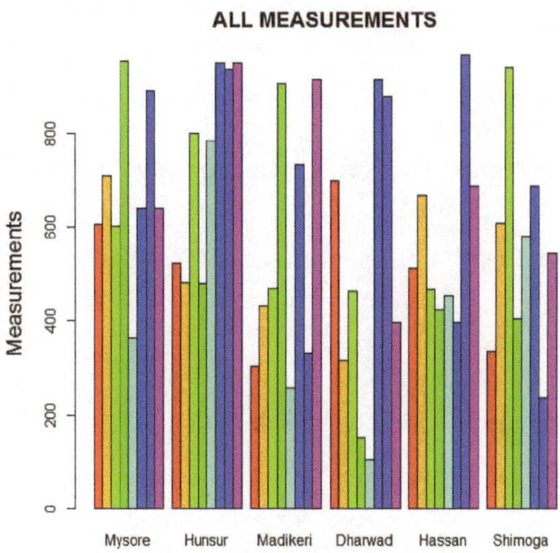

Put your text characters in between the inverted commas to make your own text labels "" and issue the command

**my.data <- read.csv("C:/researchrdata/ 6_col_data.csv")**

**my.dat**

**attch(my.dat)**

**barplot(as.matrix(my.dat), main= "ALL MEASUREMENTS", ylab = "Measurements", cex.lab = 1.5, cex.main = 1.4, beside=FALSE, col=rainbow(8))**

When you change beside = FALSE, then the chart will become a stacked histogram.

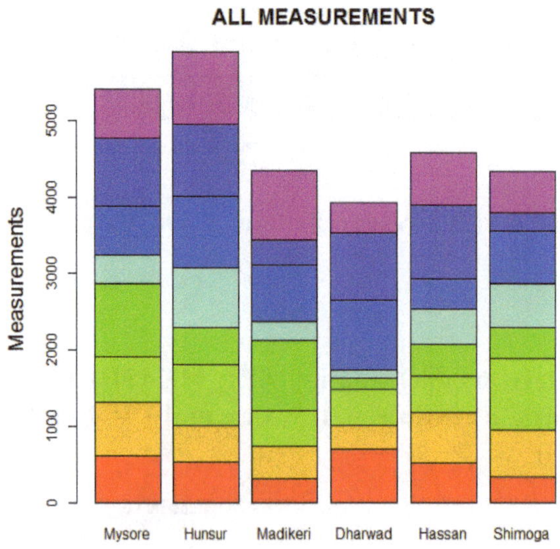


You can plot the points and a straight line to show fit, but with one column of the data only will be shown

**my.data <- read.csv("C:/researchrdata/ 6_col_data.csv")**

**my.dat**

**attch(my.dat)**

**qqnorm(Mysore,pch =16, col=rainbow(8))**

**qqline(Mysore,col = "red")**

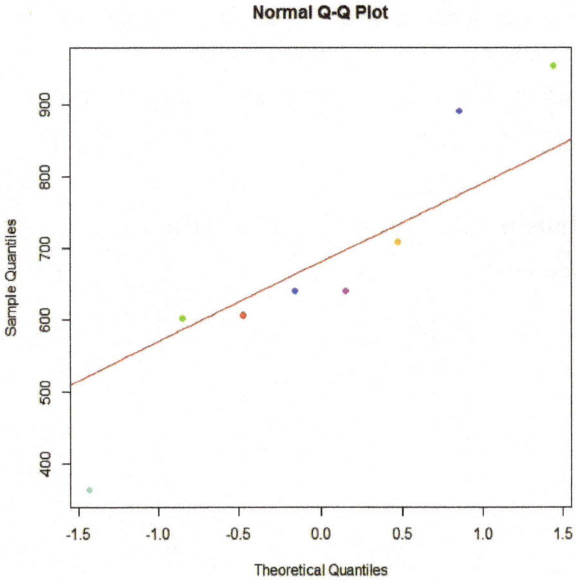


To see the distribution of the data use a Stripchart

**my.data <- read.csv("C:/researchrdata/ 6_col_data.csv")**

**my.dat**

**attch(my.dat)**

**stripchart(my.dat,pch =16, col=rainbow(8))**

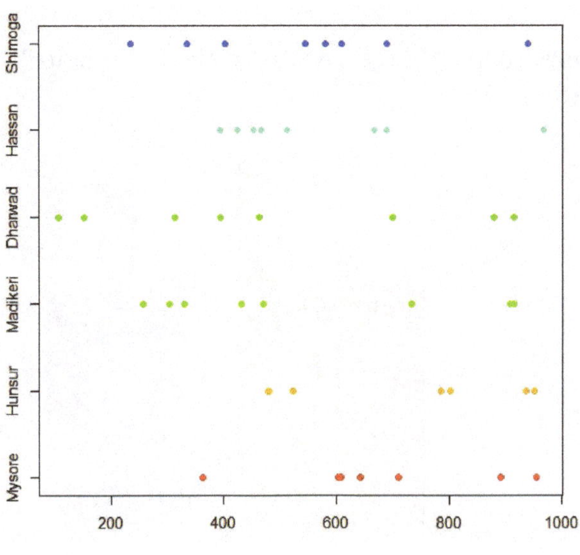

To see the correlation between two sets of observations (Mysore and Hunsur) and to Compare 2 columns numerically

**my.data <- read.csv("C:/researchrdata/ 6_col_data.csv")**

**my.dat**

**attch(my.dat)**

**lot(Mysore, Hunsur,pch =16, col=rainbow(8))**

**cor(Mysore, Hunsur)**

[1] -0.2104854

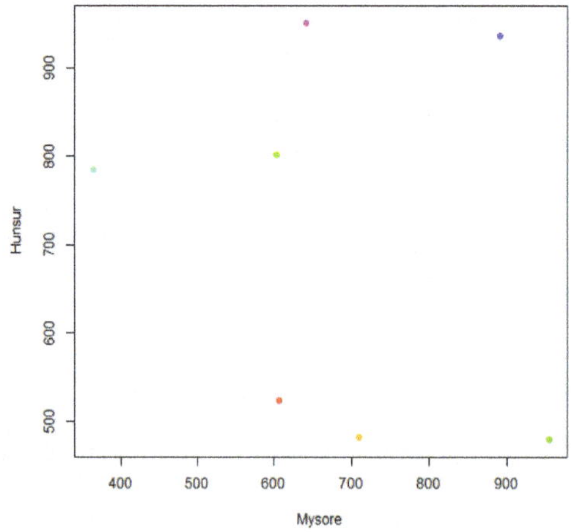


To draw barplots

**my.data <- read.csv("C:/researchrdata/ 6_col_data.csv")**

**my.dat**

**attch(my.dat)**

**barplot(Mysore, main="Place", xlab="Height of plants", col=rainbow(8))**

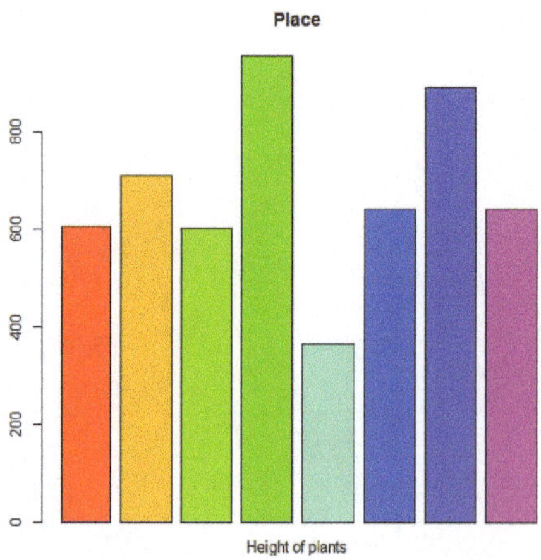

**my.data <- read.csv("C:/researchrdata/ 6_col_data.csv")**

**my.dat**

**attach(my.dat)**

**barplot(Mysore, main="Place", xlab="Height of plants",, col=rainbow(8), horiz=TRUE)**

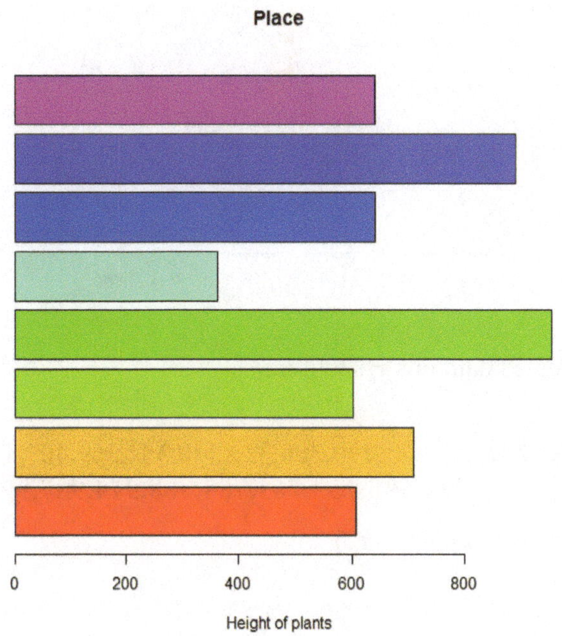


Bagplot wherein the bag contains 50% of all points. The bivariate median would be approximated. The fence separated points in the fence from points outside. Outliers are also displayed. For this you first load a package using the command library (package_name)

**library(aplpack)**

**my.data <-read.csv("C:/researchrdata/6_col_data.csv")**

**my.dat**

**attch(my.dat)**

**bagplot(Mysore,Hunsur, xlab="Height of plants", ylab="Places", main="Bagplot Example")**

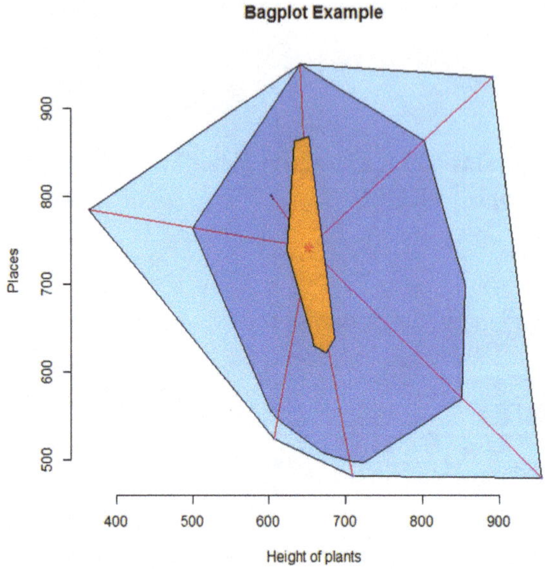


Drawing the collected data in a Histogram

**nd.x<-**
**c(113,117,235,238,252,263,271,290,300,321,340,354,369,407,417,427,436, 465,48 4,494,609,613,622,696,753,763,788,888,974,987,999)**

**hist(nd.x, breaks=15, col="red")**

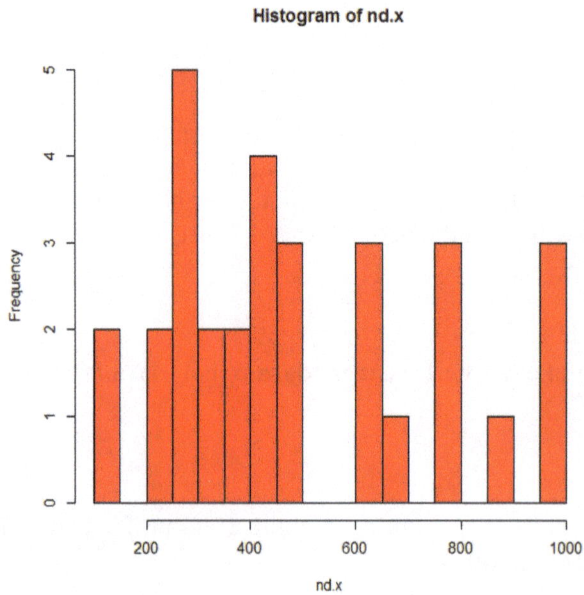

Drawing the data in a Histogram overlaid with normal distribution curve

```
nd.x<-
c(113,117,235,238,252,263,271,290,300,321,340,354,369,407,417,427,436,4
65,484,494,609,613,622,696,753,763,788,888,974,987,999)

h<-hist(nd.x, breaks=10, density=10, col= rainbow(12),
xlab="Accuracy", main="Overall")

xfit<-seq(min(nd.x),max(nd.x),length=40)

yfit<-dnorm(xfit,mean=mean(nd.x),sd=sd(nd.x))

yfit <- yfit*diff(h$mids[1:2])*length(nd.x)

lines(xfit, yfit, col= "black", lwd=2)
```

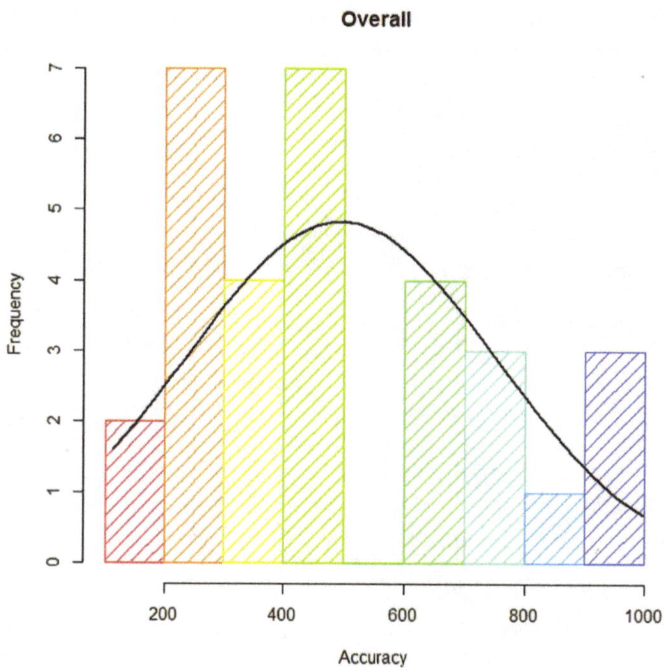


Drawing Multiple graphs in same page

```
# 2x2 4 graphs in a page

par(mfrow=c(2,2))

my.data <- read.csv("C:/researchrdata/ 6_col_data.csv")

my.dat
```

```
attch(my.dat)
plot(Mysore, Hunsur, main="Scatterplot of Mysore and Hunsur")
plot(Hunsur, Madikeri, main="Scatterplot of Hunsur and Madikeri")
hist(Mysore, main= "Histogram of Mysore")
boxplot(Madikeri, main="Boxplot of Madikeri")
```

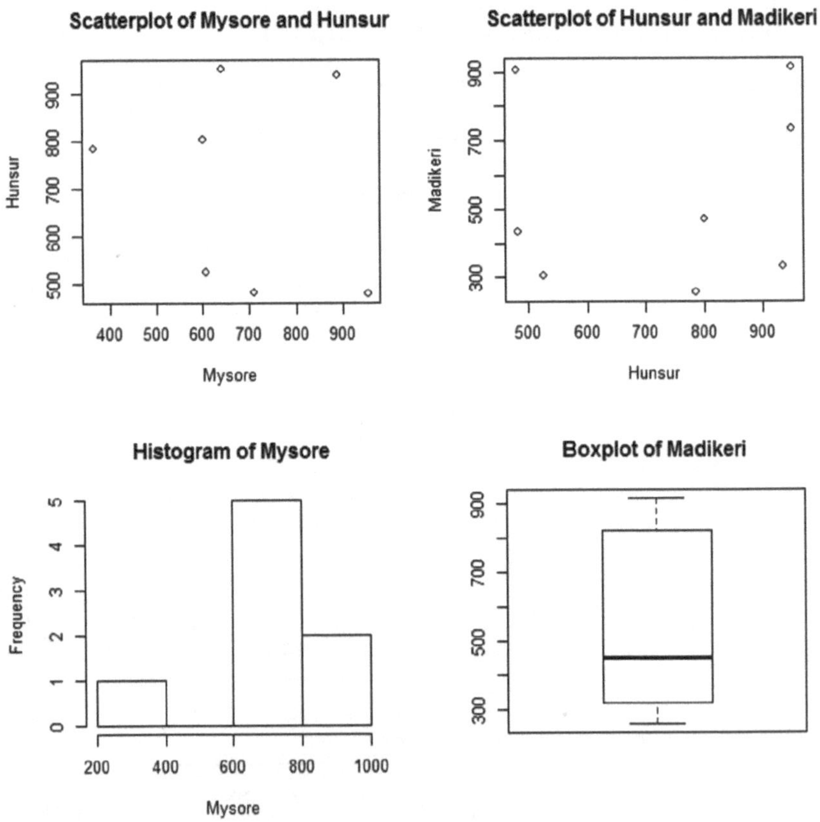


```
# 4 figures arranged in 2 rows and 2 columns
par(mfrow=c(2,2))
par(mfrow=c(2,2))
my.data <- read.csv("C:/researchrdata/ 6_col_data.csv")
my.dat
attch(my.dat)
```

plot(Mysore, Hunsur, pch=16, col= rainbow(12), main= "Scatterplot of Mysore and Hunsur")

plot(Hunsur, Madikeri, pch=16, col= rainbow(12),main="Scatterplot of Hunsur and Madikeri")

hist(Mysore,col= rainbow(12), main="Histogram of Mysore")

boxplot(Madikeri, col= rainbow(12), main="Boxplot of Madikeri")

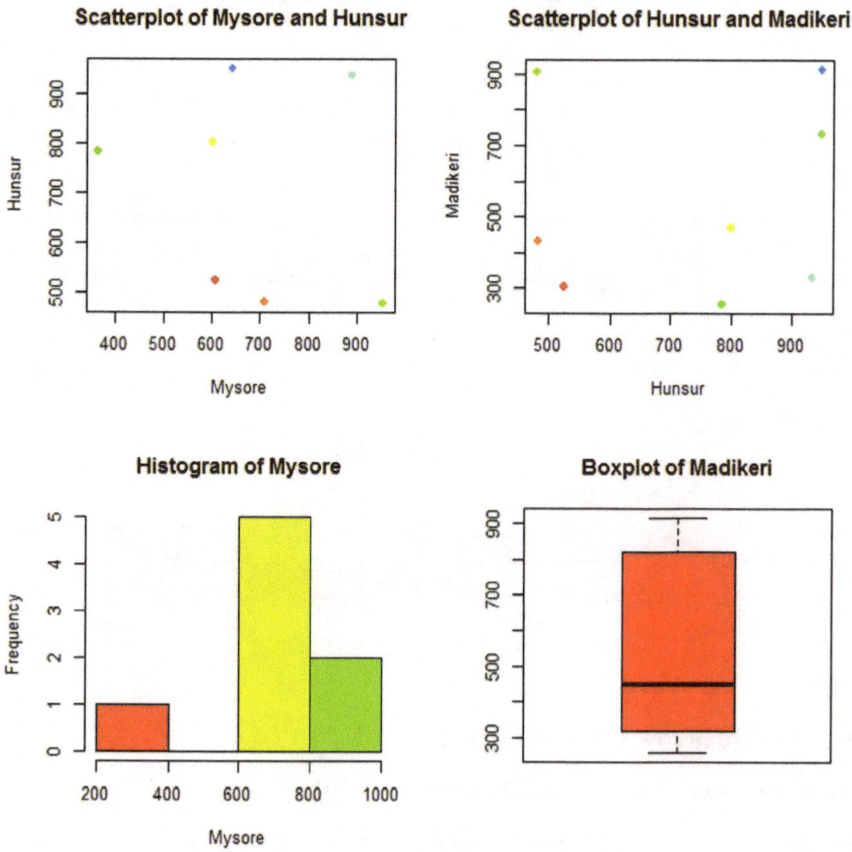


# 3 figures arranged in 3 rows and 1 column

**par(mfrow=c(3,1))**

**my.data <- read.csv("C:/researchrdata/ 6_col_data.csv")**

**my.dat**

**attch(my.dat)**

```
hist(Mysore,col= topo.colors (8), main="Histogram of Mysore")
hist(Hunsur,col=terrain.colors(8), main="Histogram of Hunsur")
hist(Madikeri,col= heat.colors(8), main="Histogram of Madikeri")
```

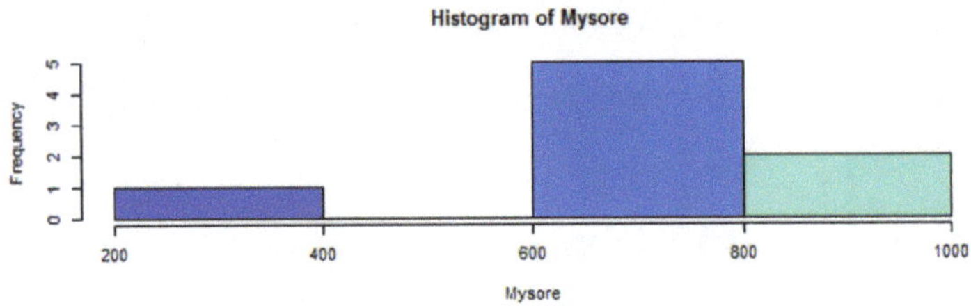


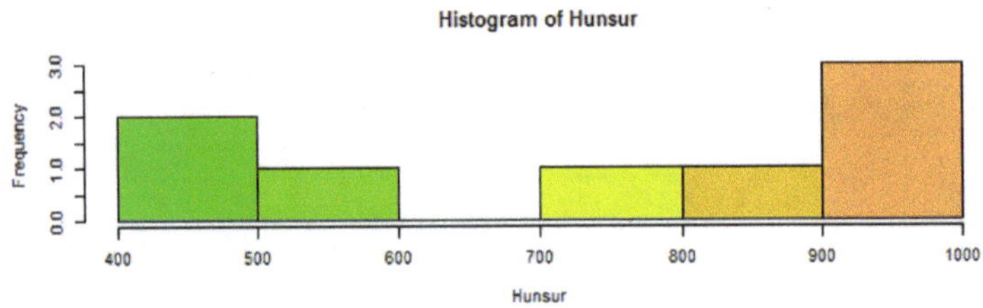


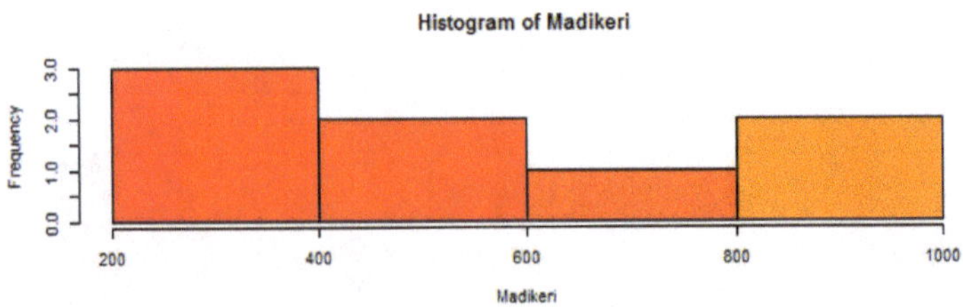


```
# One figure in row 1 and two figures in row 2
layout(matrix(c(1,1,2,3),2,2,byrow=TRUE))
my.data<-read.csv("C:/researchrdata/ 6_col_data.csv")
my.dat
attch(my.dat)
hist(Mysore,col=topo.colors (8), main="Histogram of Mysore")
hist(Hunsur,col=terrain.colors(8), main="Histogram of Hunsur")
hist(Madikeri,col=heat.colors(8), main="Histogram of Madikeri")
```

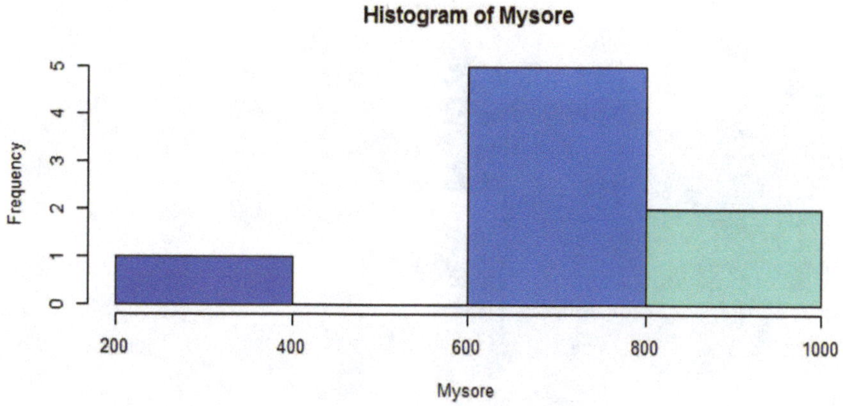


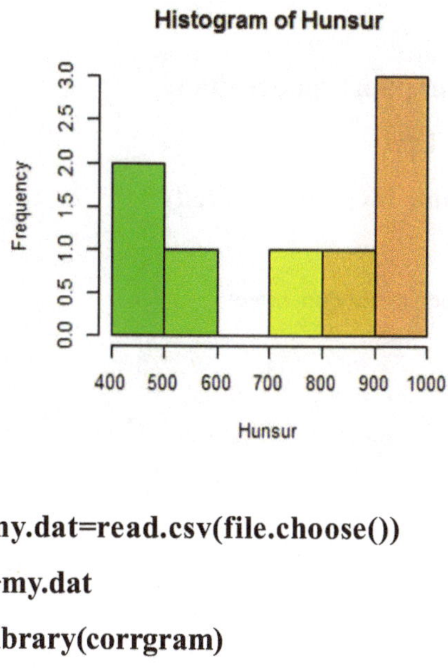


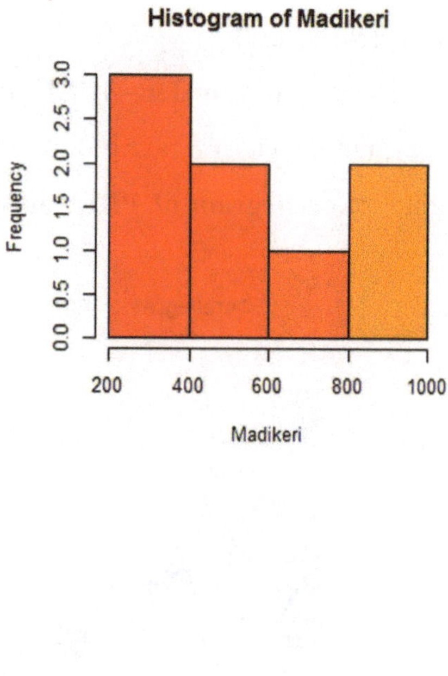


```
my.dat=read.csv(file.choose())
>my.dat
library(corrgram)
corrgram(my.dat,order=TRUE, lower.panel=panel.shade,
upper.panel=panel.pie,text.panel=panel.txt,
main= "Corellegram of PRSV infection factors in PC2/PC1 Order")
```

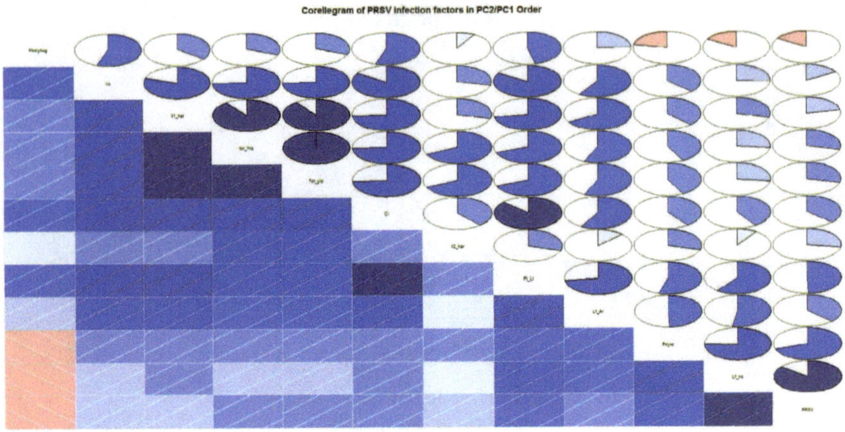


```
library(corrgram)
corrgram(my.dat, order=NULL, lower.panel=panel.shade,
upper.panel=NULL, text.panel=panel.txt,
main= "Corellegram of PRSV infection factors(unsorted)")
```

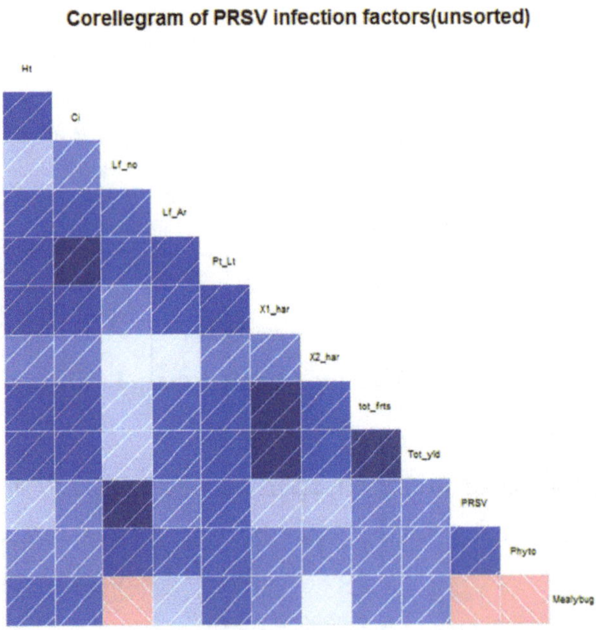

library(corrplot)

M <-cor(my.dat)

corrplot.mixed(M, addtextlabel= “lt”, diag= “u”)

corrplot.mixed(M,col=terrain.colors(10),cl.length=10, addtextlabel=“lt”,diag= “u”)

corrplot.mixed(M,col=terrain.colors(15), cl.length=15, addtextlabel=“lt”, diag= “u”)

corrplot.mixed(M,col=rainbow(15), cl.length=15, addtextlabel= “lt”, diag=“u”)

col1<-colorRampPalette(c(“#7F0000“,”red“,”#FF7F00“,”green“,”gray“,”cyan “,“#007FFF “,”blue“,”#00007F”))

corrplot.mixed(M,col=col1(9))

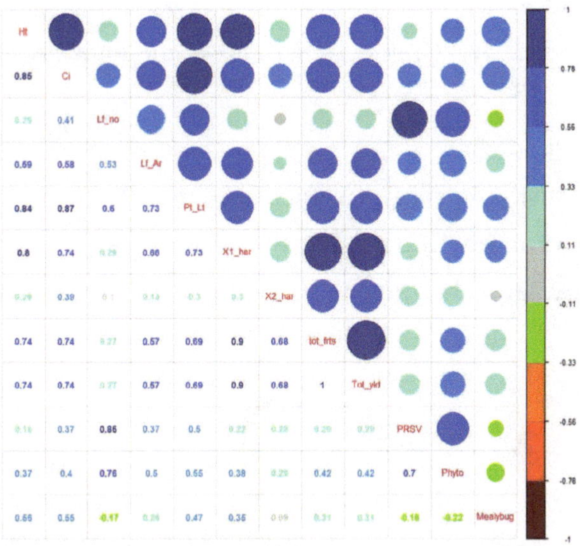


\# Kernel Density Plot

**my.data <- read.csv(“C:/researchrdata/ 6_col_data.csv”)**

**my.dat**

**attch(my.dat)**

**d.my.dat <- density(Mysore) # returns the density data**

**plot(d.my.dat, col= “blue”, lty = 3, lwd = 3)# plots the results**

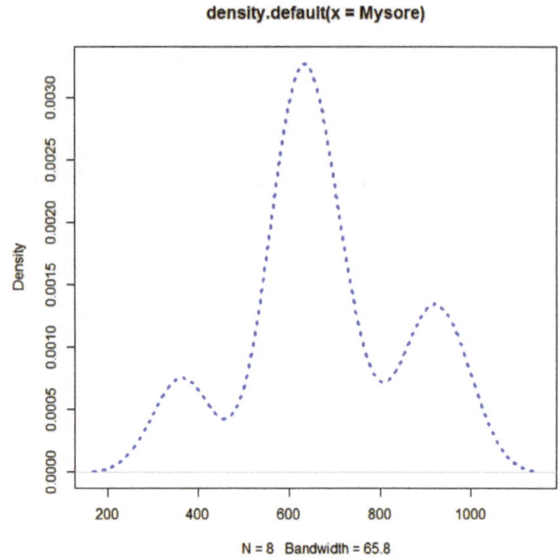


```
# Filled Density Plot
my.data <- read.csv("C:/researchrdata/ 6_col_data.csv")
my.dat
attch(my.dat)
d.my.dat<-density(Mysore) # returns the density data
plot(d.my.dat, main="Kernel Density of Mysore")
polygon(d.my.dat, col= "green", border="blue", lty = 3, lwd = 3)
```

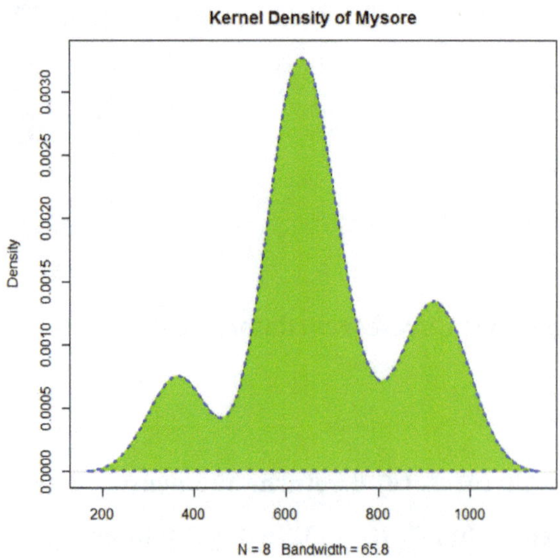

**# dotchart but only one colum of data will be drawn**
**my.data <- read.csv("C:/researchrdata/ 6_col_data.csv")**
**my.dat**
**attch(my.dat)**
**dotchart(Mysore, pch = 16,col=rainbow(8))**

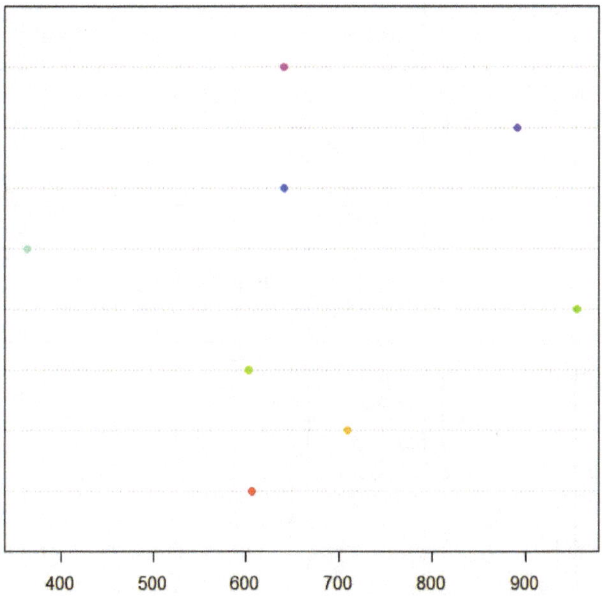


pairs command will draw a scatter plot of a matrix or a data frame, at all combinations.

**my.data <- read.csv("C:/researchrdata/6_col_data.csv")**

**my.dat**

**attch(my.dat)**

**pairs(my.dat,pch =16, col=rainbow(8))**

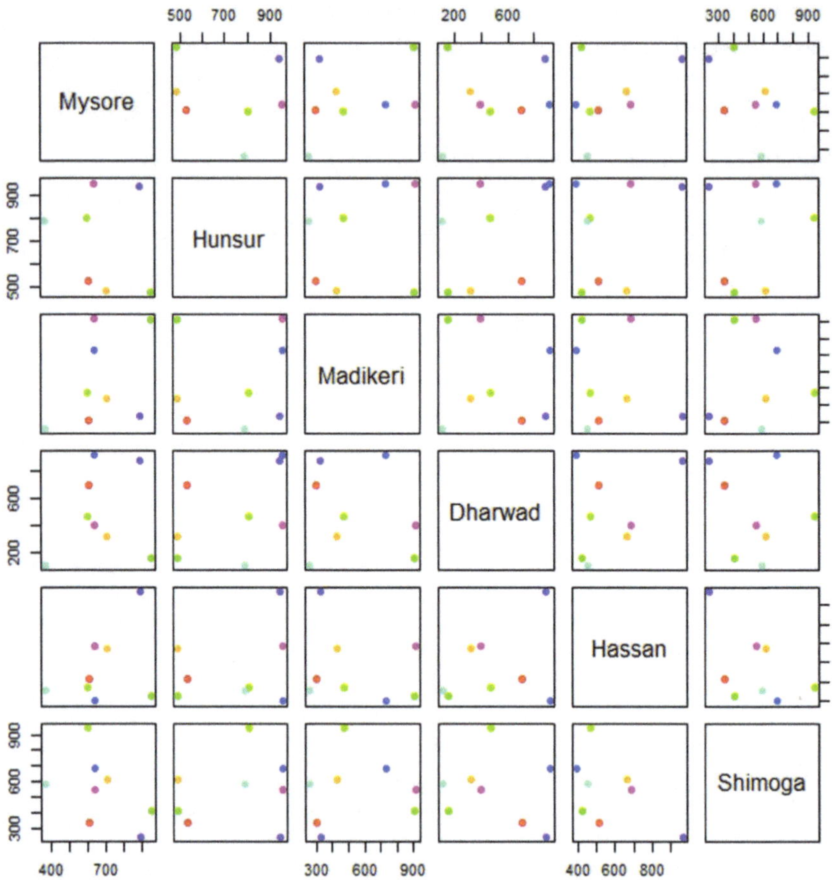


```
library(lattice)
my.data <- read.csv(“C:/researchrdata/ 6_col_data.csv”)
my.dat
attch(my.dat)
splom(my.dat, pch = 16, col = rainbow(8))
```

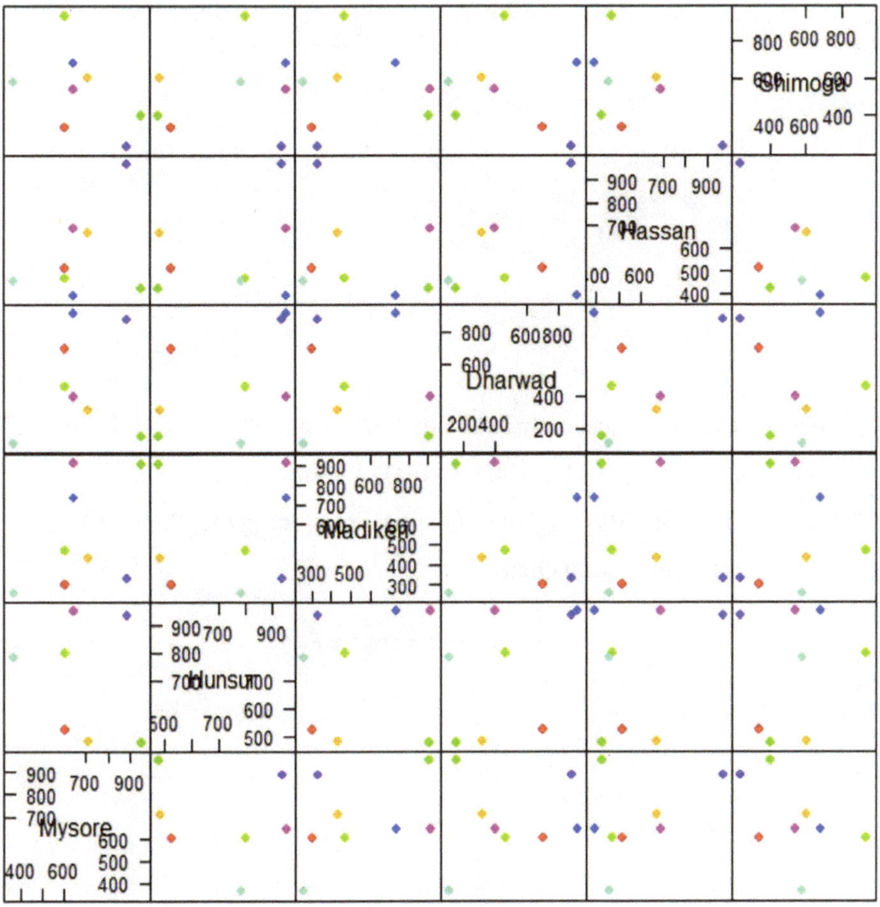


Scatter Plot Matrix

USING GGPLOT 2

```
library(ggplot2)
w <- read.csv(file=”C:/researchrdata/small_set_1.csv”, head=TRUE, sep=“,”)
p <- ggplot(data=w, aes(x=Plant, y=Height, color=Area))
p + layer(geom=”point”, geom_params=list(size=4))
```

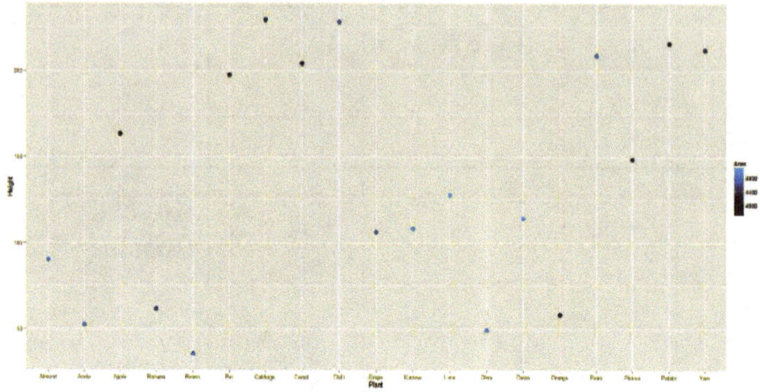

```
w <- read.csv(file="C:/researchrdata/small_set_1.csv", head=TRUE, sep=",")
p <- ggplot(data=w, aes(x=Height, y=Area)) + geom_jitter()
p + facet_grid(. ~ Quality)+ ggtitle("Quality Fractioning the Plants") +
xlab("Height") + ylab("total Quality")
```

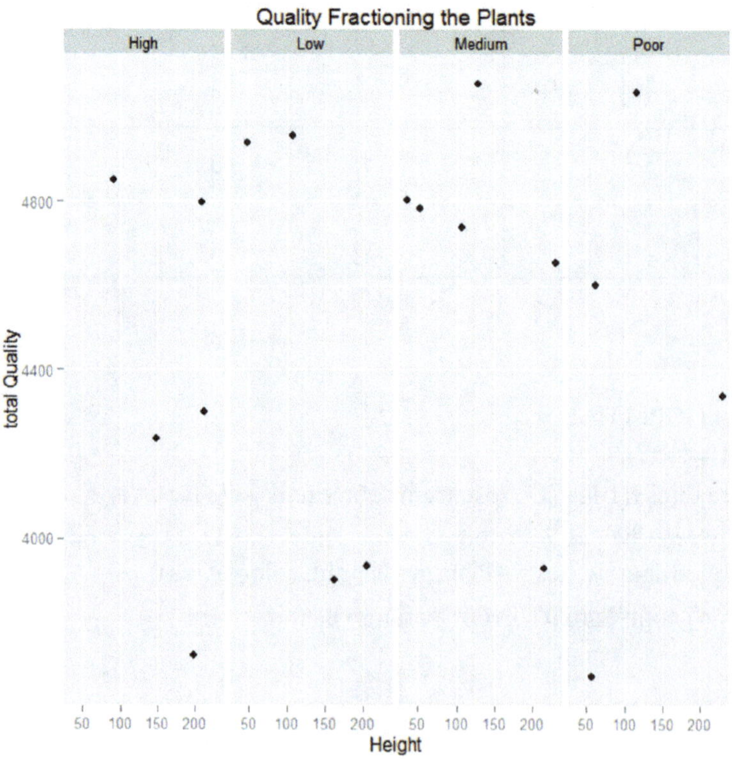

```
w <- read.csv(file="C:/researchrdata/small_set_1.csv", head=TRUE, sep=",")
p <- ggplot(data=w, aes(x=Height, y=Area)) + geom_jitter()
p + facet_grid(Quality ~ .)+ ggtitle("Quality Fractioning the Plants") +
xlab("Height") + ylab("total Quality")
```

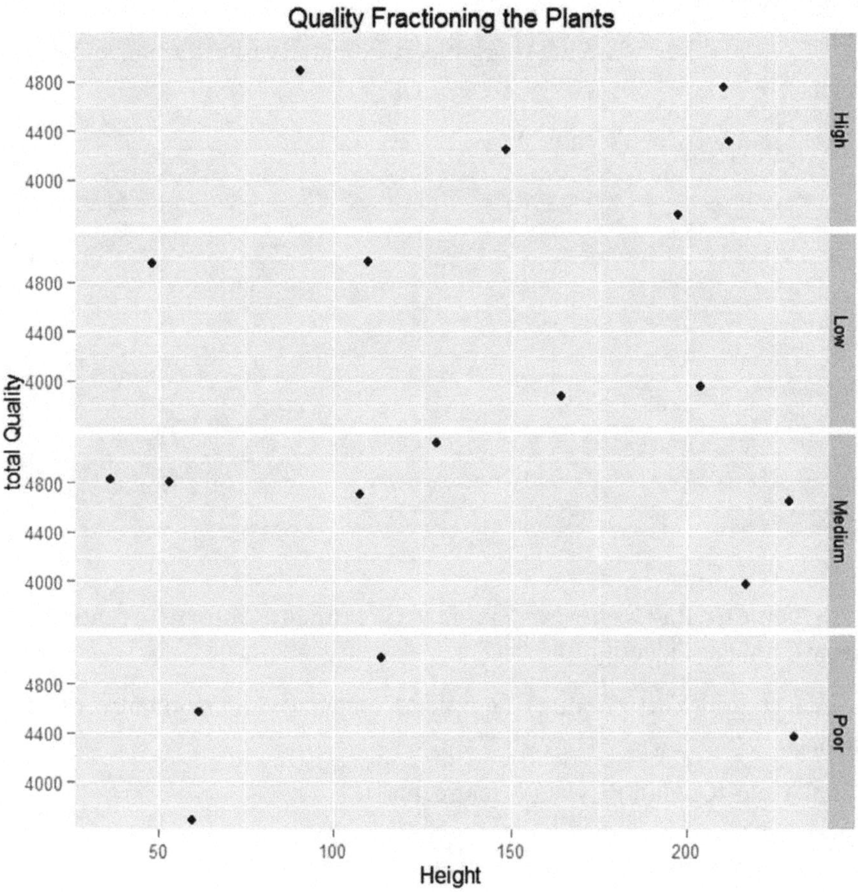

```
library (ggplot2)
w <- read.csv(file= "C:/Users/dksamuel_now/Desktop/
R-book/data/multi_aicrp.csv", head=TRUE, sep=",")

attach(w)

p <- ggplot(data=w, aes(x=Yield, y=Treatment)) +
geom_jitter()

◆Centre)+ ggtitle("Comparative Performance of Treat-
ments") +

xlab("Yield") + ylab("Centre")
```

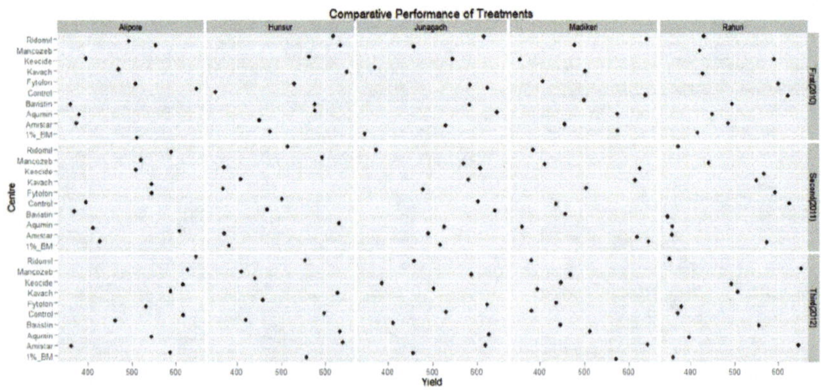


```
p <- ggplot(data=w, aes(x=Yield, y=Treatment, color = Treatment))+
geom_bar(stat="identity")

p + facet_grid(Year◆Centre)+ ggtitle("Comparative Performance of Treat-
ments") +

xlab("Yield(Kg/Ha)") + ylab("Treatments")
```

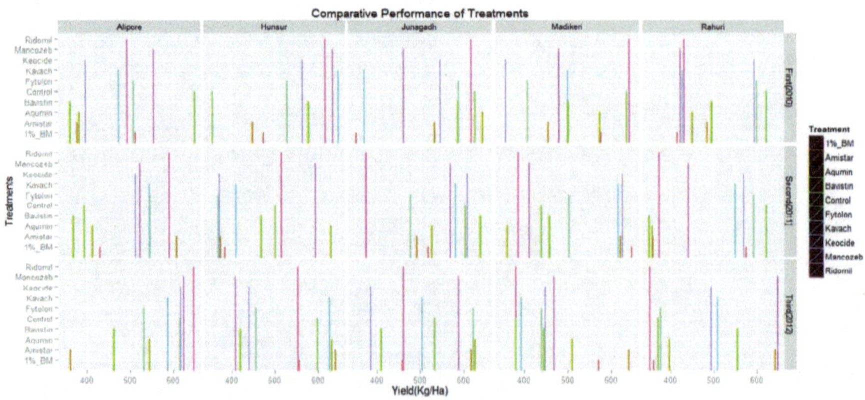

p <- ggplot(data=w, aes(x=Yield, y=Year, color = Year))+ geom_bar (stat=”identity”)

p + facet_grid(Treatment ~ Centre)+ theme(legend.position=”none”) + ggtitle(“Comparative Performance of Treatments”) +

xlab(“Yield(Kg/Ha)”) + ylab(“Year”)

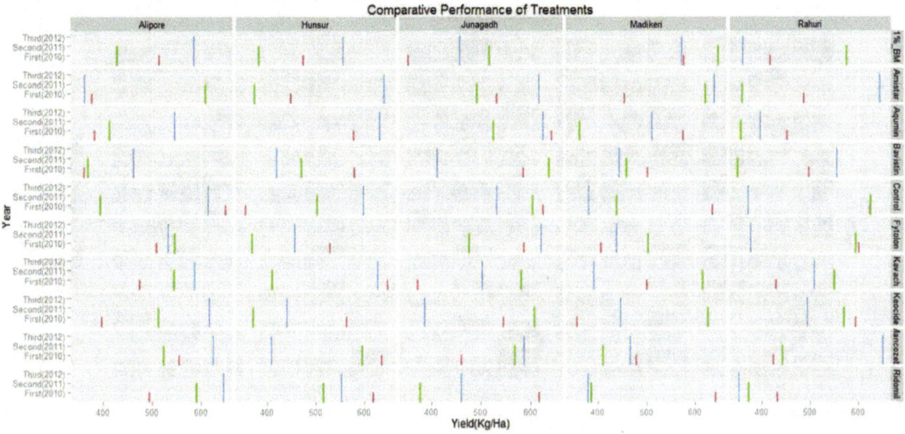


p <- ggplot(data=w, aes(x=Yield, y=Year, color = Treatment))+ geom_bar(stat=”identity”)

p + facet_grid(Treatment ~ Centre)+ ggtitle(“Comparative Performance of Treatments”) +

xlab(“Yield(Kg/Ha)”) + ylab(“Performance over the Time Period”)

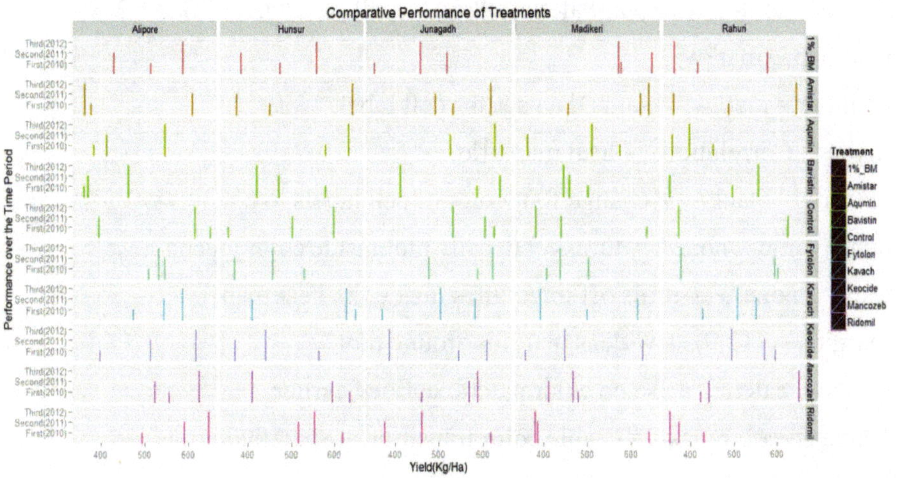

```
p <- ggplot(data=w, aes(x=Year, y=Yield, color = Treatment))+ geom_bar(stat=
"identity")

p + facet_grid(Treatment�Centre)+ ggtitle("Comparative Performance of Treat-
ments") +

xlab("Performance over the Time Period") + ylab("Yield(Kg/Ha")
```

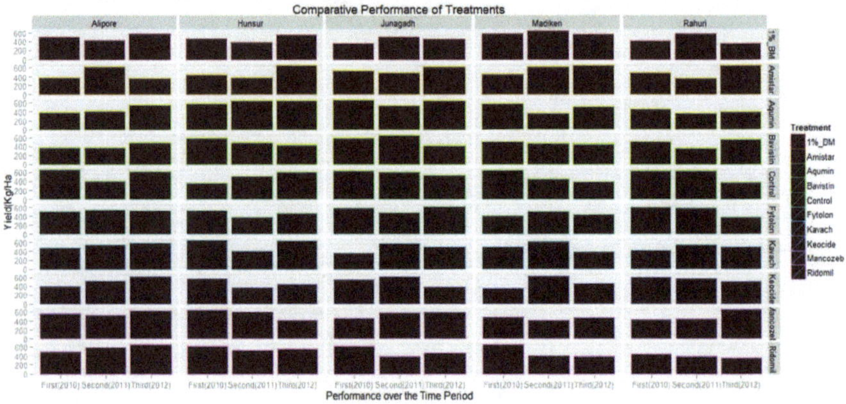

```
w <- read.csv(file="C:/researchrdata/heat_ map.csv",
head=TRUE, sep=",")

attach(w)

rnames <- w[,1]

mat_w <- data.matrix(w[,2:ncol(w)])

rownames(mat_w) <- rnames

my_palette <- colorRampPalette(c("red", "yellow", "green"))(n = 299)

heatmap.2(mat_w,

cellnote = mat_w, # same data set for cell labels

main = "Correlation", # heat map title

notecol= "black", # change font color of cell labels to black

density.info="none", # turns off density plot inside color legend

trace= "none", # turns off trace lines inside the heat map

margins =c(12,9), # widens margins around plot

col=my_palette, # use on color palette defined earlier

# breaks=col_breaks, # enable color transition at specified limits

dendrogram= "row", # only draw a row dendrogram
```

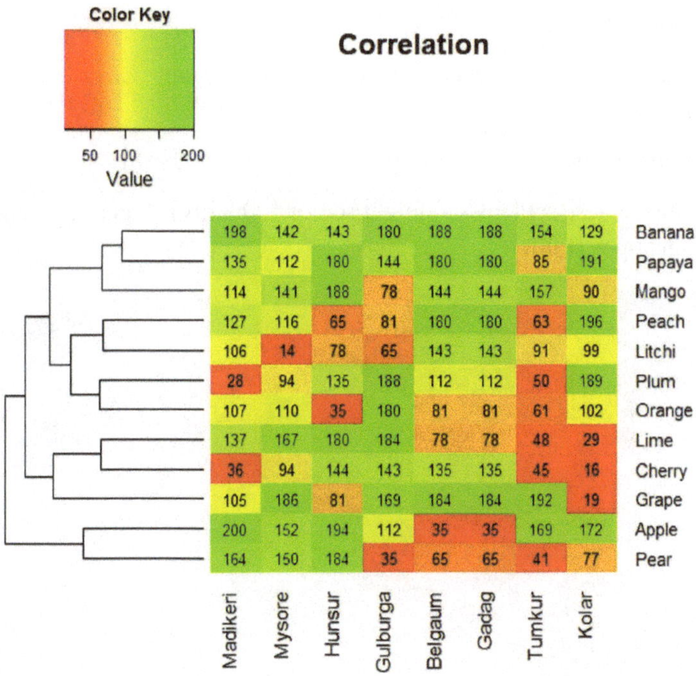


```
library(ggplot2)

w <- read.csv(file="C:/ C:/researchrdata/small_set_1.csv", head=TRUE, sep=",")

p <- ggplot(data=w, aes(x=Plant , y=Height , color = Quality))

ggplot(data=w, aes(x=Plant , y=Height, color = Quality)) + geom_line(aes(colour = Quality, group = Quality))
```

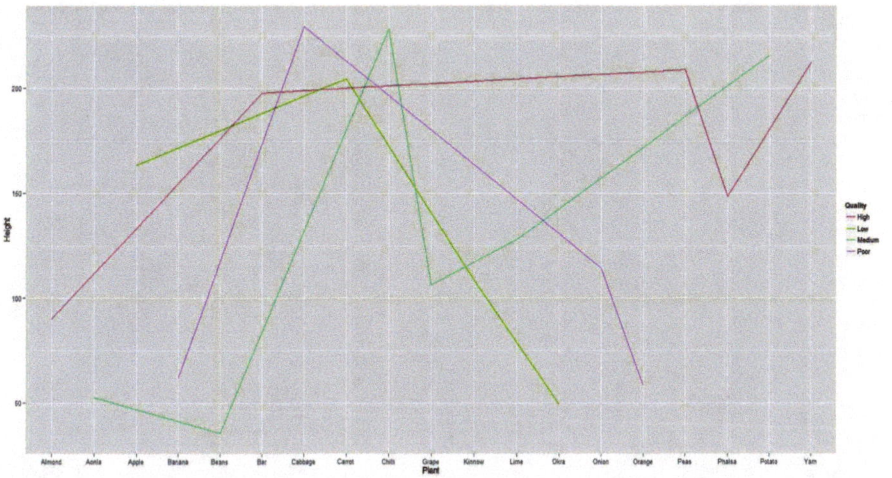

```
library(ggplot2)

w <- read.csv(file="C:/researchrdata/data/small_set_1.csv", head=TRUE, sep=",")

p <- ggplot(data=w, aes(x=Plant, y=Height, color = Quality))

ggplot(data=w, aes(x=Plant, y=Height , color = Quality)) + geom_line(aes(colour = Habit, group = Habit))
```

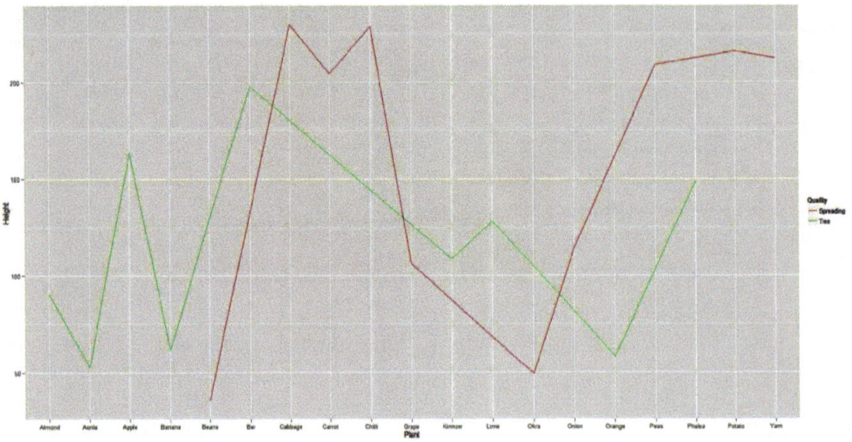

```
library(ggplot2)

w <- read.csv(file="C:/researchrdata/data/small_set_1.csv", head=TRUE, sep=",")

p <- ggplot(data=w, aes(x=Plant, y=Height, color = Quality))

ggplot(data=w, aes(x=Plant, y=Height, color = Quality)) + layer(geom="jitter")
```

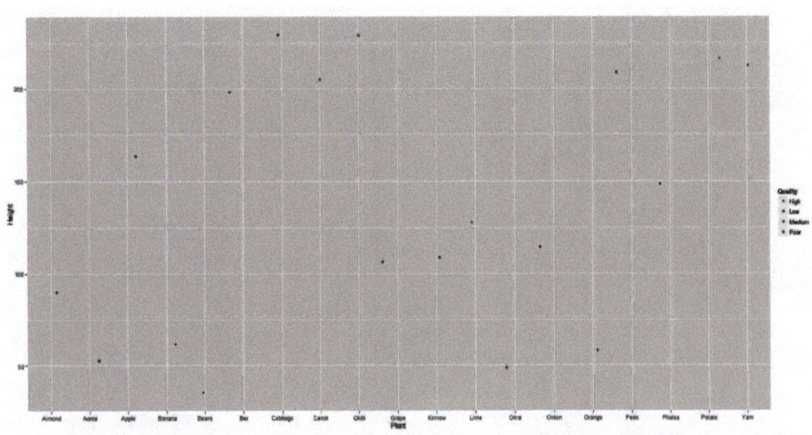

```
library(ggplot2)
w <- read.csv(file="C:/researchrdata/data/small_set_1.csv", head=TRUE, sep=",")
ggplot(data=w, aes(x=Plant, y=Height, color=Area))+ geom_point()+ theme(axis.title.y = element_text(colour="grey20",size=12,angle=90,hjust=.5,vjust=.5,face="plain"))
```

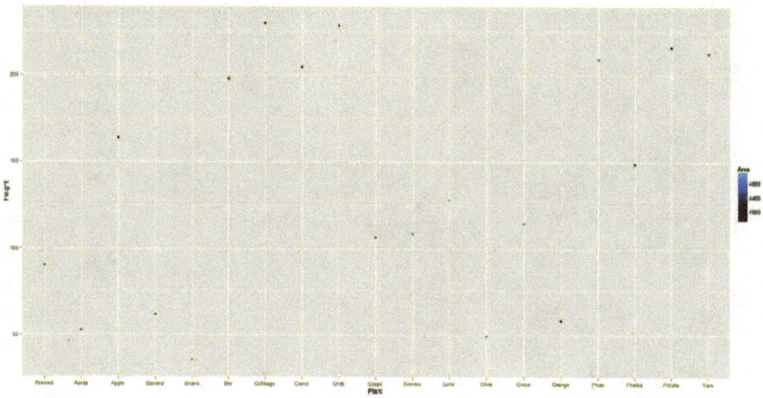

```
library(ggplot2)
w <- read.csv(file="C:/researchrdata/multiples_1.csv", head=TRUE, sep=",")
p <- ggplot(data=w, aes(x= Quality))
p + stat_bin()
p + stat_bin(geom="bar")
```

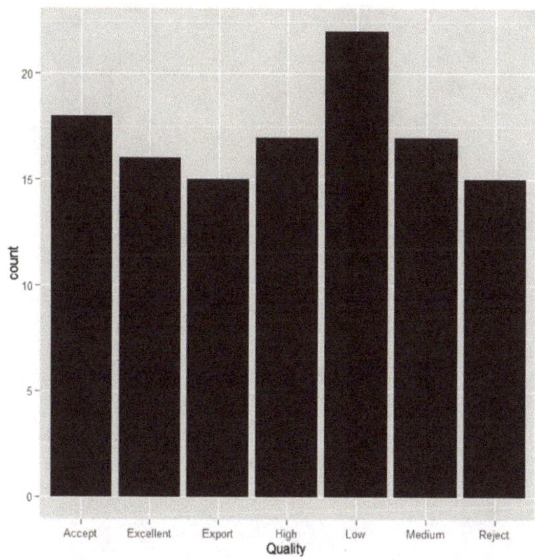

```
library(ggplot2)
w <- read.csv(file="C:/researchrdata/multiples_1.csv", head=TRUE, sep=",")
p <- ggplot(data=w, aes(x= Quality))
p + stat_bin()
p + stat_bin(geom="point", size=5)
```

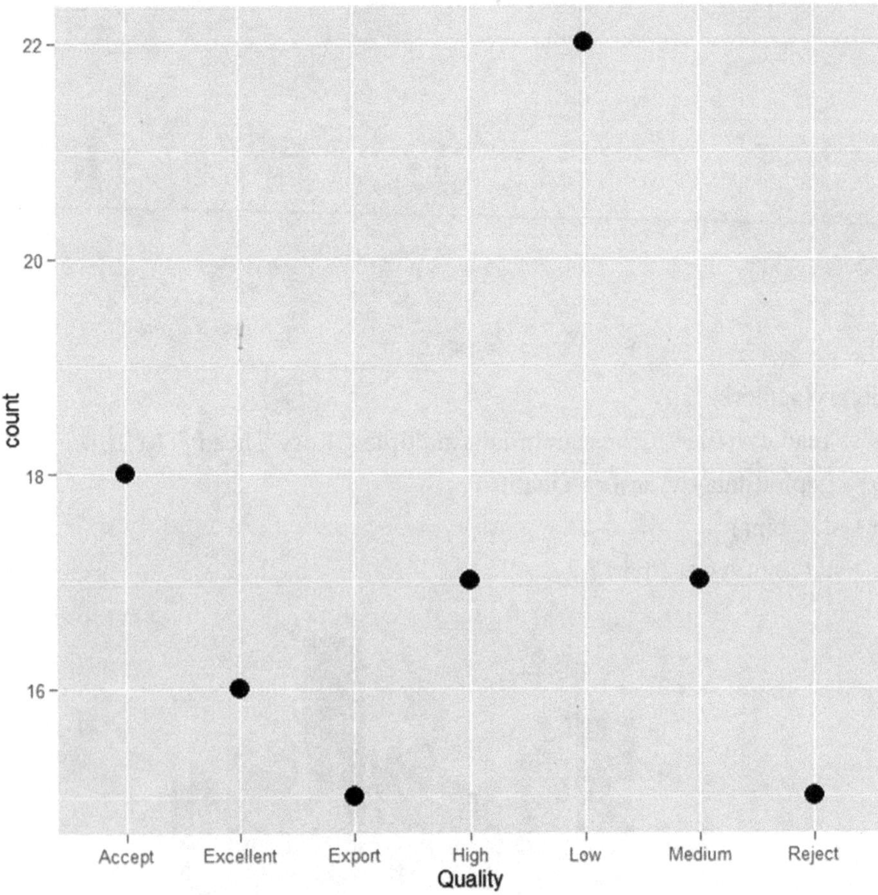

```
library(ggplot2)
w <- read.csv(file="C:/researchrdata/multiples_1.csv", head=TRUE, sep=",")
p <- ggplot(data=w, aes(x= Quality))
p + stat_bin()
p + stat_bin(geom="tile")
```

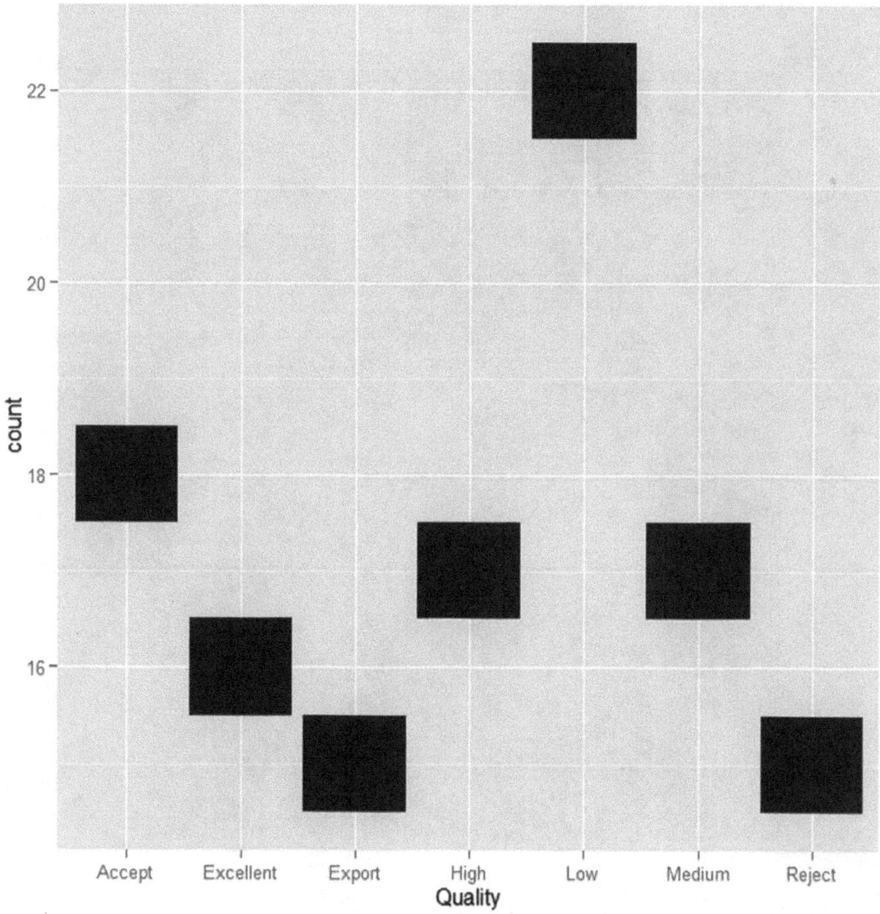

```
library(ggplot2)
w <- read.csv(file= "C:/researchrdata/multiples_1.csv", head=TRUE, sep=",")
p <- ggplot(data=w, aes(x= Quality))
p + stat_bin()
p <- ggplot(w, aes(x=factor(Place), fill=Quality))
p + geom_bar()
```

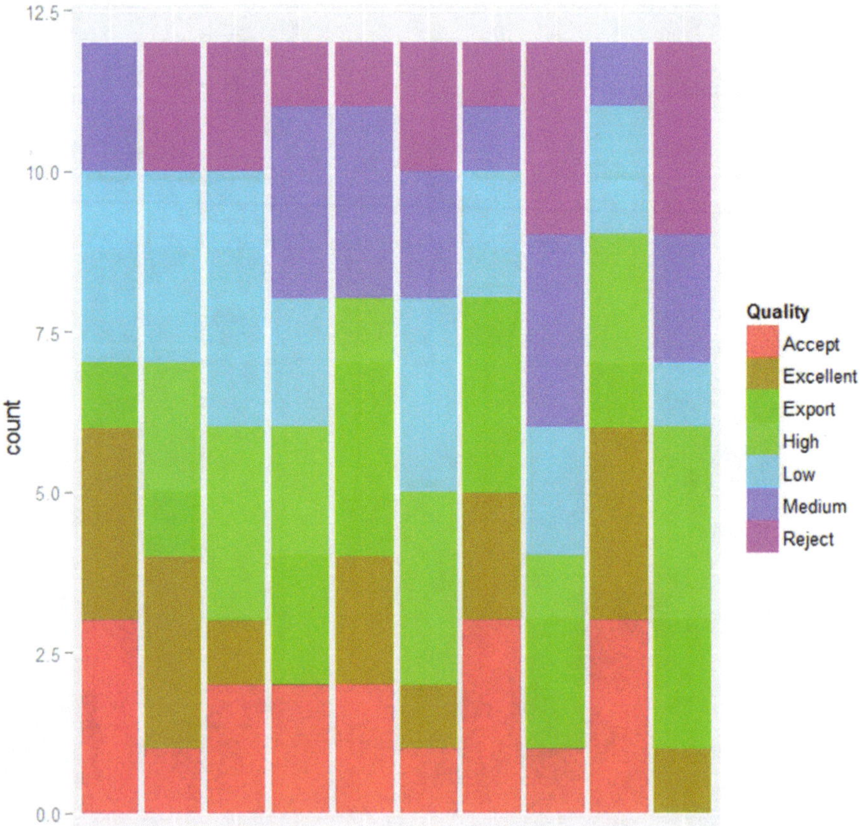


Guwahati Hunsur Imphal Jhansi Madras Madurai Nagpur Ranchi Tiruchi Tumkur
Factor (Place)

```
library(ggplot2)
w <- read.csv(file="C:/researchrdata/multiples_1.csv", head=TRUE, sep=",")
p <- ggplot(data=w, aes(x= Quality))
p + stat_bin()
p + geom_bar() + coord_flip()
```

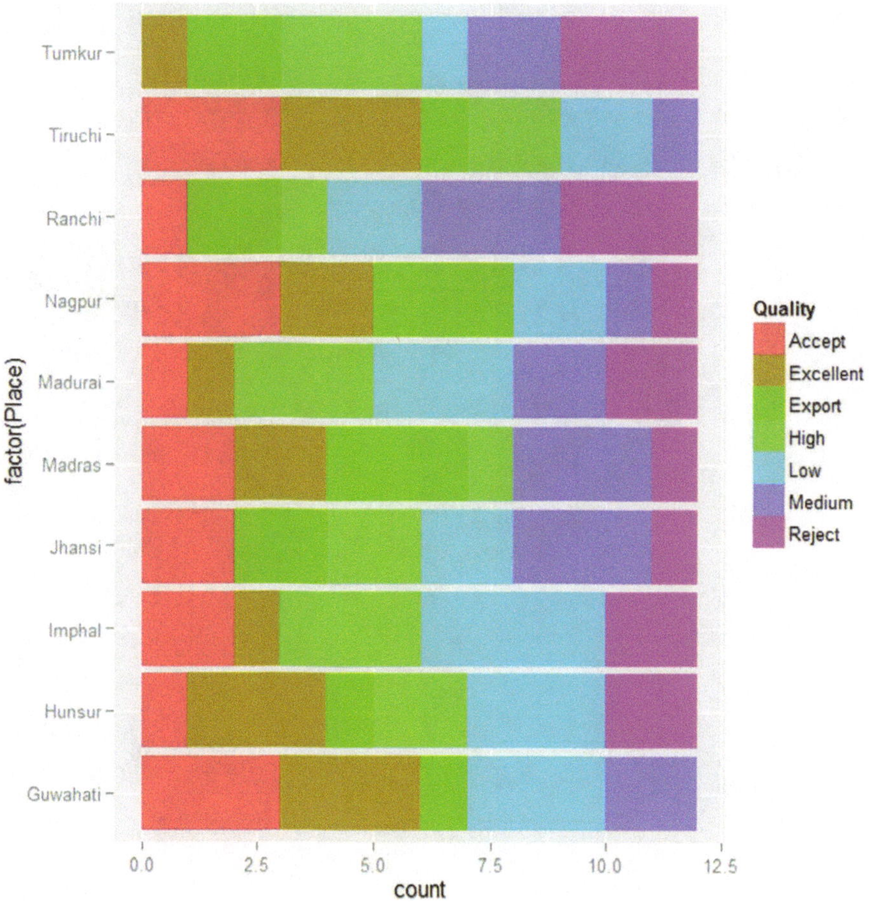

```
library(ggplot2)
w <- read.csv(file= "C:/researchrdata/multiples_1.csv", head=TRUE, sep=",")
p <- ggplot(data=w, aes(x= Quality))
p + stat_bin()
p + geom_bar() + coord_polar(theta="y")
```

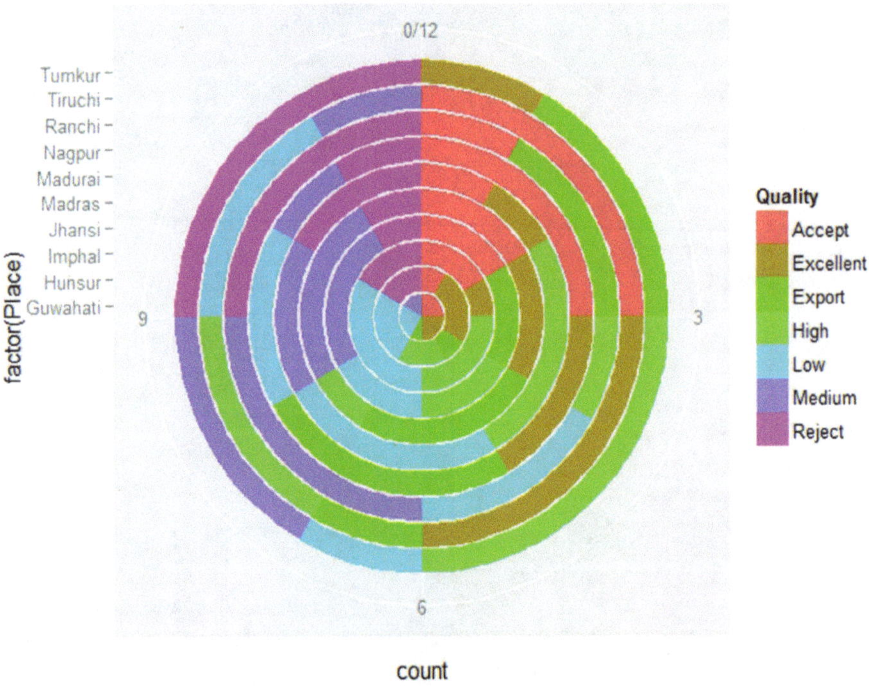

```
library(ggplot2)
w <- read.csv(file="C:/researchrdata/multiples_1.csv", head=TRUE, sep= ",")
p <- ggplot(data=w, aes(x= Quality))
p + stat_bin()
a <- ggplot(data=w, aes(x= Fruits, y =Place))
a <- a + geom_point(size = 5)
a <- a + facet_wrap(♦Place)
a <- a + xlab("Number of Fruits") + ylab("Places of Collection") + ggtitle("Weight of g Fruit Collections from Places")
a
```

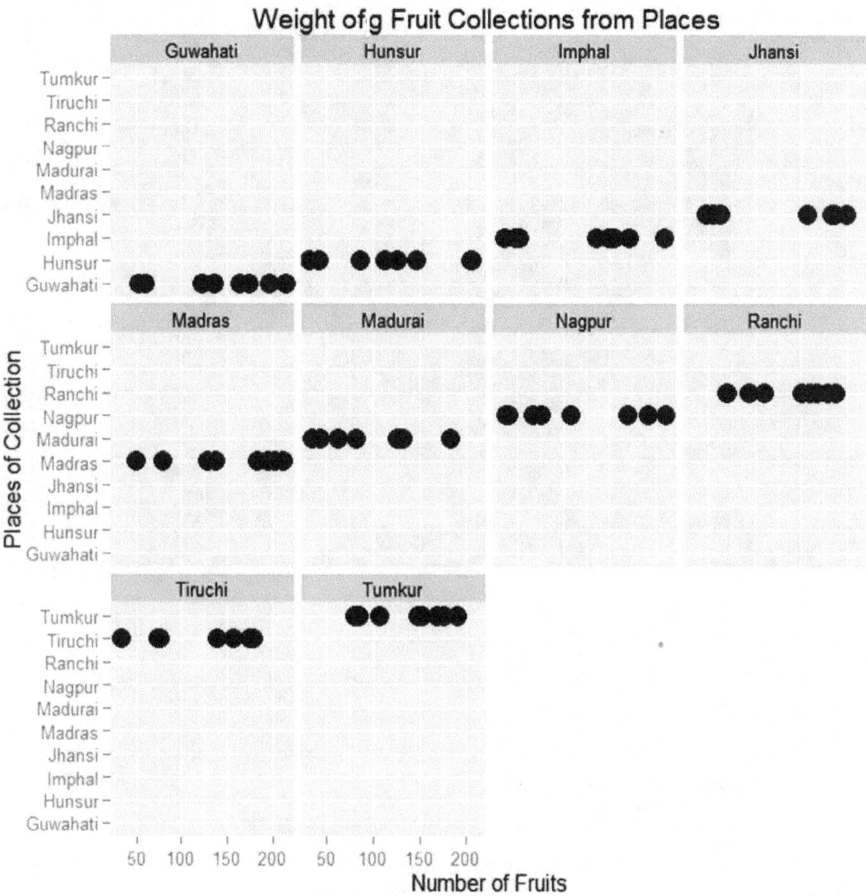

```
library(ggplot2)
w <- read.csv(file= "C:/researchrdata/multiples_1.csv", head=TRUE, sep=",")
p <- ggplot(data=w, aes(x= Quality))
p + stat_bin()
a <- ggplot(data=w, aes(x= Fruits, y =Place))
a <- a + geom_point(size = 5)
                    ♦Quality)
a <- a + xlab("Number of Fruits") + ylab("Distribution of Fruits") +
ggtitle("Relative Fruit Numbers")
a
```

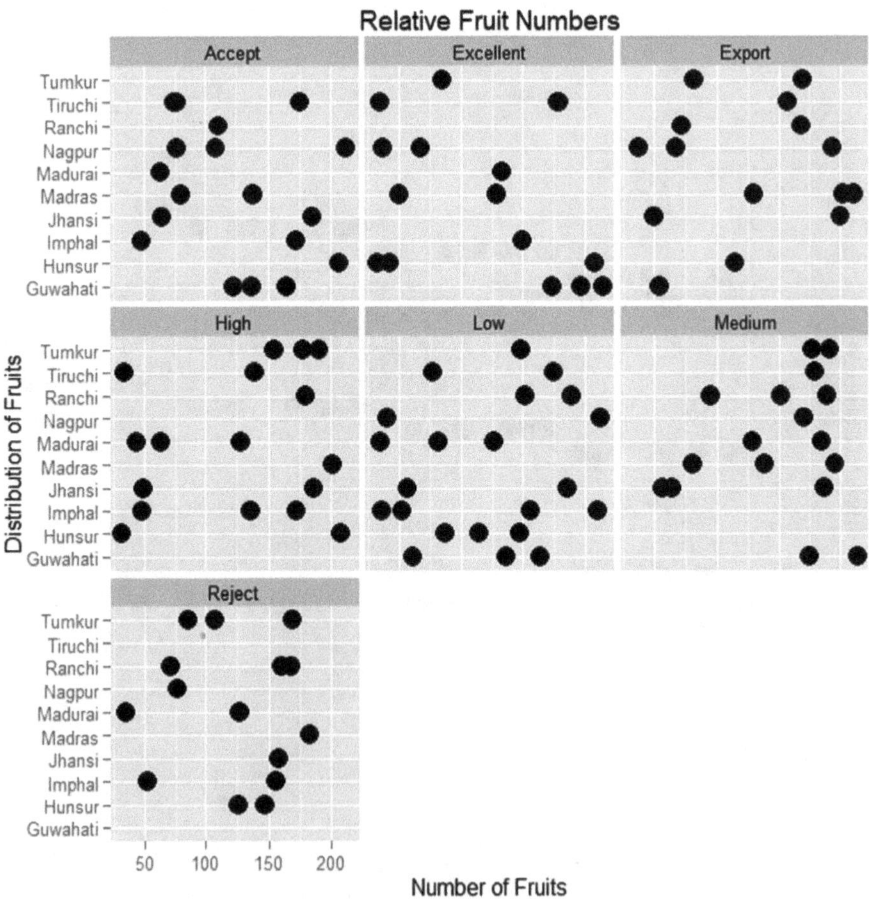

```
library(ggplot2)
w <- read.csv(file="C:/researchrdata/multiples_1.csv", head=TRUE, sep=",")
p <- ggplot(data=w, aes(x= Quality))
p + stat_bin()
a <- ggplot(data=w, aes(x= Fruits, y =Place))
a <- a + geom_point(size = 2)
a <- a + facet_grid(Area ~ Quality)
a <- a + xlab("Number of Fruits") + ylab("Distribution of Fruits") + ggtitle("Relative Fruit Numbers")
a
```

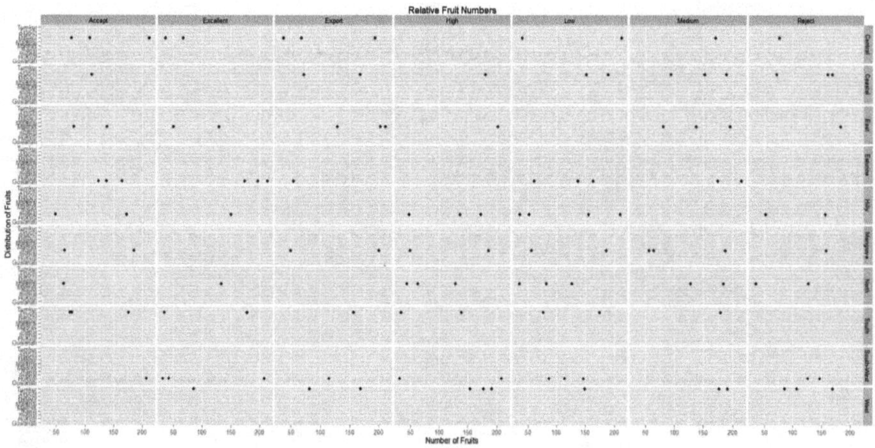

```
library(ggplot2)
w <- read.csv(file= “C:/researchrdata/multiples_1.csv”, head=TRUE, sep= “,”)
p <- ggplot(data=w, aes(x= Quality))
p + stat_bin()
a <- ggplot(data=w, aes(x= Fruits, y =Place))
a <- a + geom_point(size = 2)
a <- a + facet_grid(Area ~ Quality)
a <- a + xlab(“Number of Fruits”) + ylab(“Distribution of Fruits”) + ggtitle(“Relative Fruit Numbers”)
a
```

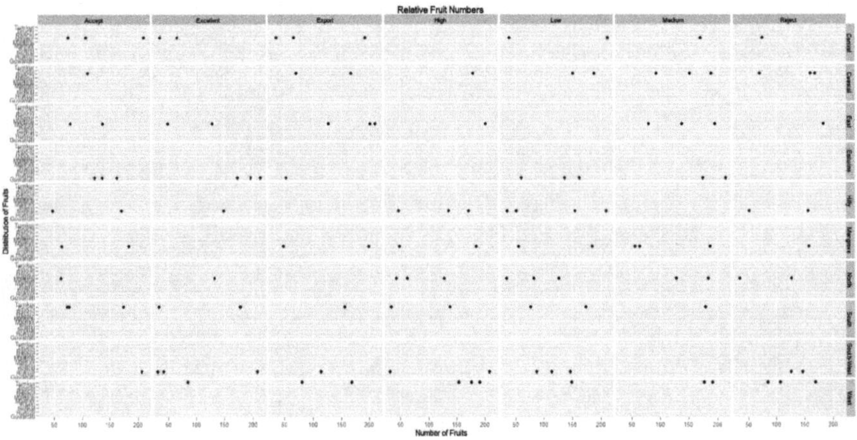

```
library(ggplot2)
w <- read.csv(file="C:/researchrdata/multiples_1.csv", head=TRUE, sep=",")
p <- ggplot(data=w, aes(x= Quality))
qplot(Fruits,Flowers, data=w, geom=c("point", "smooth"))
```

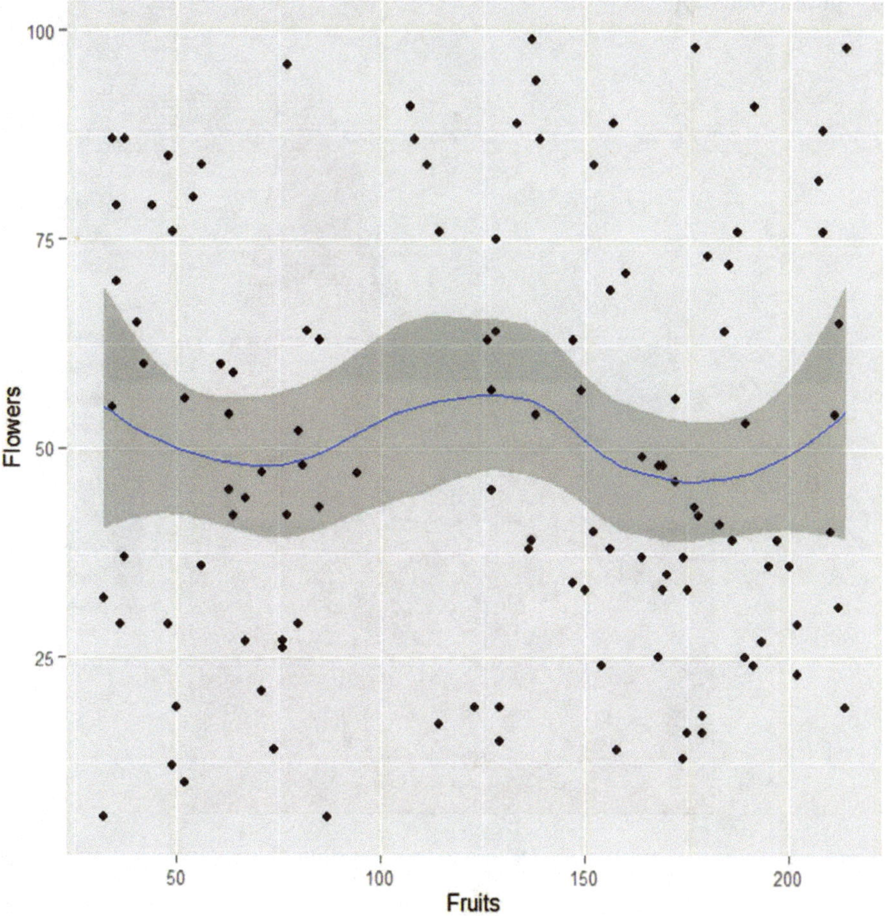

```
library(ggplot2)
w <- read.csv(file="C:/researchrdata/multiples_1.csv", head=TRUE, sep= ",")
p <- ggplot(data=w, aes(x= Quality))
p <- ggplot(w, aes(x = Fruits, fill= Place))
p + geom_density()
```

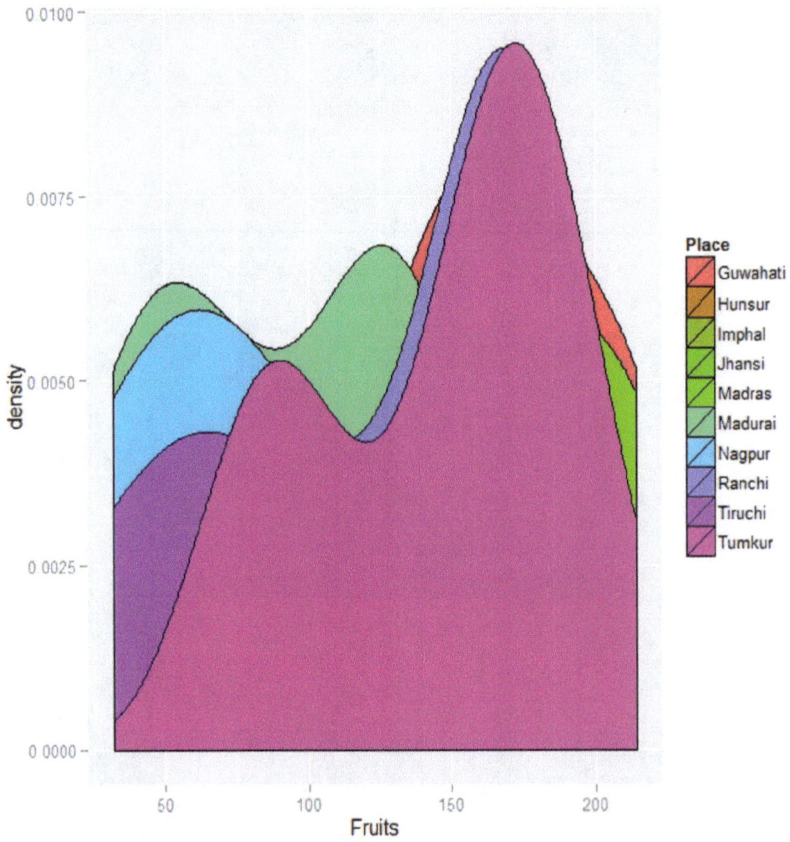

```
library(ggplot2)
w <- read.csv(file=" C:/researchrdata/multiples_1.csv", head=TRUE, sep=",")
m <- lm(Fruits ~ Yield, data=w)
mf <- fortify(m)
p <- ggplot(data=mf, aes(x=.fitted,y=.resid))
p + geom_point() +
geom_hline(y = 0) +
geom_smooth(se = FALSE)
```

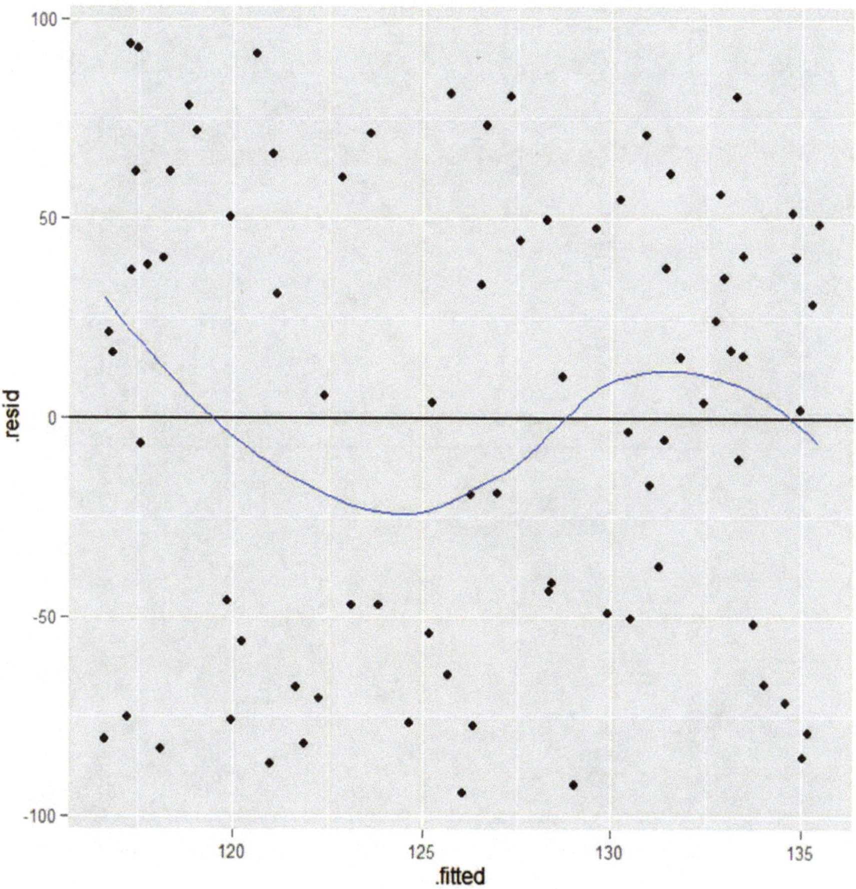

```
library(ggplot2)
w <- read.csv(file= "C:/researchrdata/multiples_1.csv", head=TRUE, sep=",")
p <- ggplot(data=w, aes(x= Quality))
p <- ggplot(data=mf, aes(x=.fitted,y=.stdresid))
p + geom_point() +
geom_hline(y=0) +
geom_hline(y=2, linetype="dashed") +
geom_hline(y=-2, linetype="dashed") +
geom_smooth(se = FALSE)
```

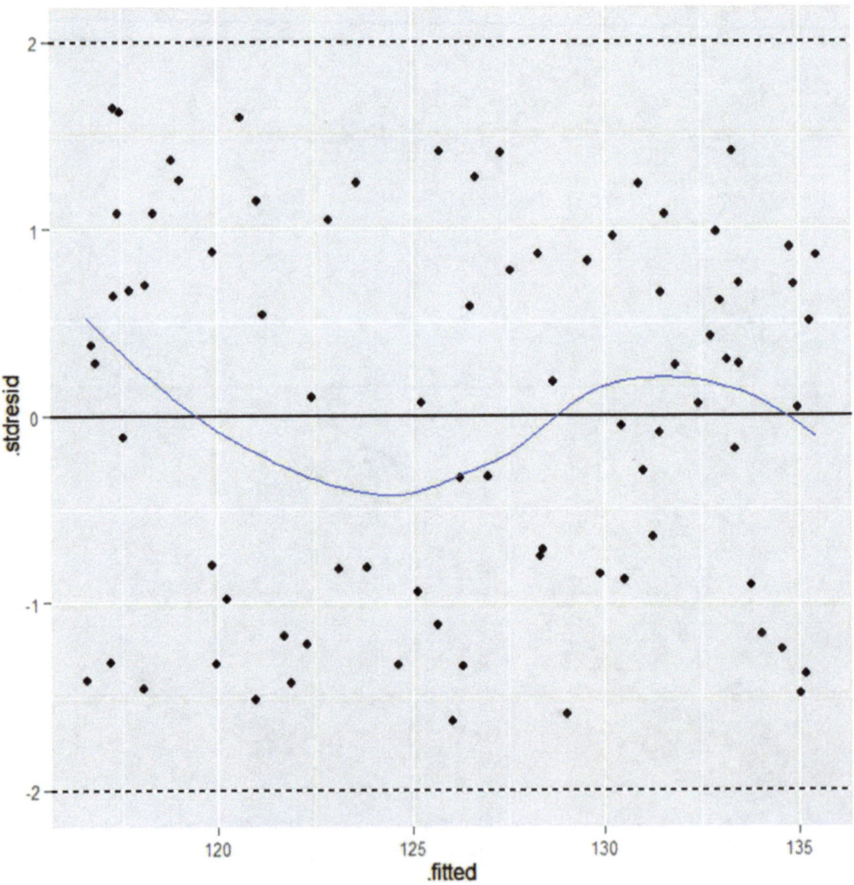

```
library(ggplot2)
w <- read.csv(file= "C:/researchrdata/multiples_1.csv", head=TRUE, sep=",")
p <- ggplot(data=w, aes(x= Quality))
wf <-fortify(m,w)
p <- ggplot(data=wf, aes(x=.fitted, y=.stdresid))
p + geom_point(aes(color=Area)) +
geom_hline(y = 0) +
geom_hline(y=2, linetype= "dashed") + geom_hline(y=-2, linetype= "dashed") +
geom_smooth(se = FALSE)
```

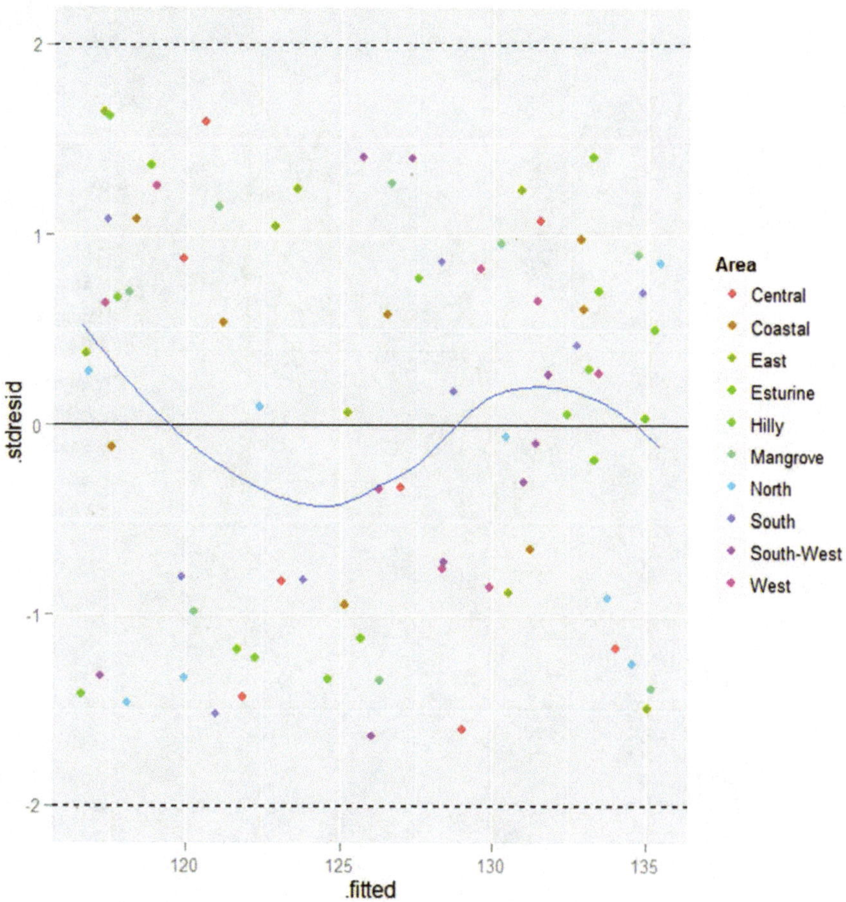

**library(ggplot2)**

**w <- read.csv(file= "C:/researchrdata/multiples_1.csv", head=TRUE, sep=",")**

**p <- ggplot(data=w, aes(x= Quality))**

**p <- ggplot(data=wf, aes(x=.fitted, y=.stdresid))**

p + geom_line(aes(color=Area)) +

geom_hline(y = 0) +

geom_hline(y=2, linetype="dashed") + geom_hline(y=-2, linetype= "dashed")

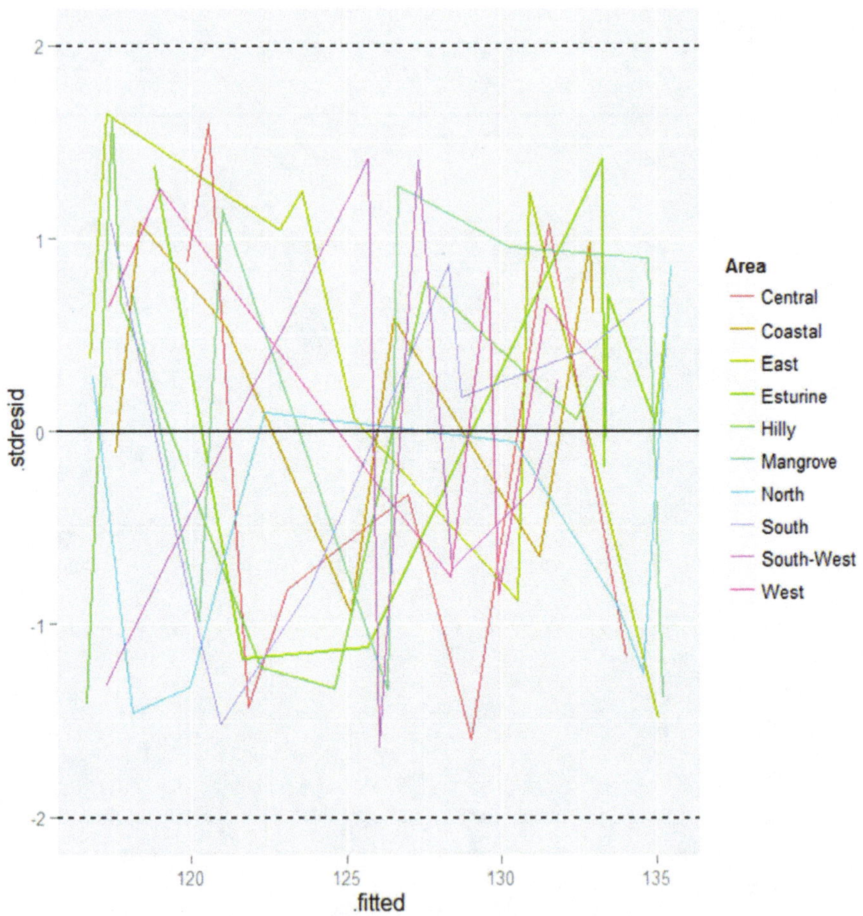

```
library(ggplot2)
library(GGally)
my.data <- read.csv("C:/researchrdata/ 6_col_data.csv")
my.dat
attch(my.dat)
ggpairs(my.dat, lower=list(continuous = "smooth"))
```

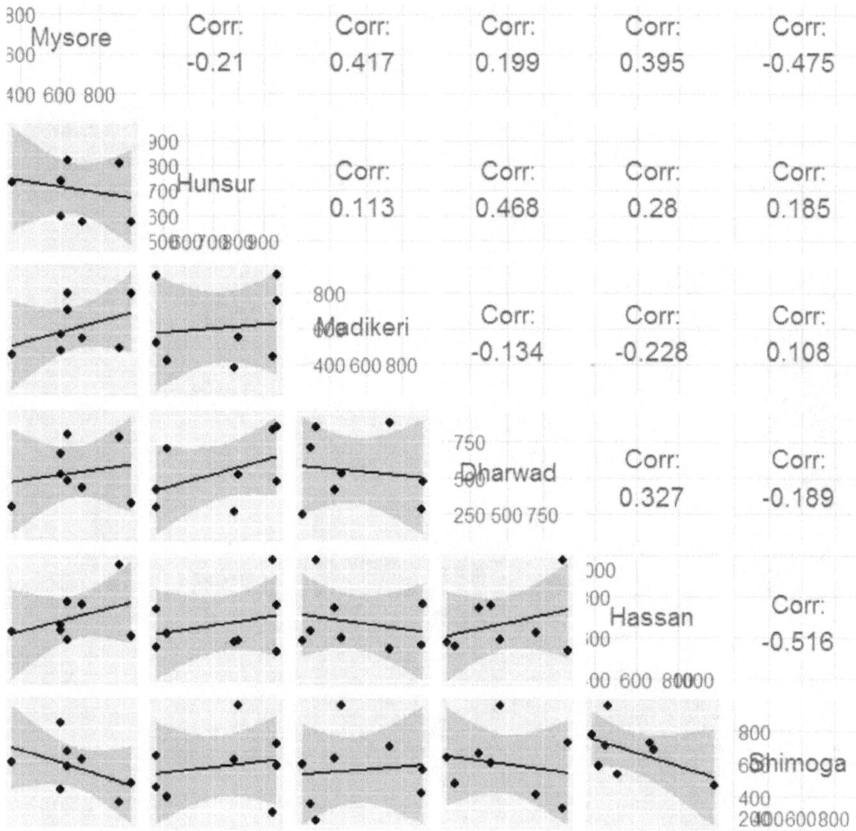

**library(ggplot2)**

**library(GGally)**

**my.data <- read.csv("C:/researchrdata/ 6_col_data.csv")**

**my.dat**

**attch(my.dat)**

**ggpairs(data=my.dat, # data.frame with variables + title="Place data", # title of the plot**

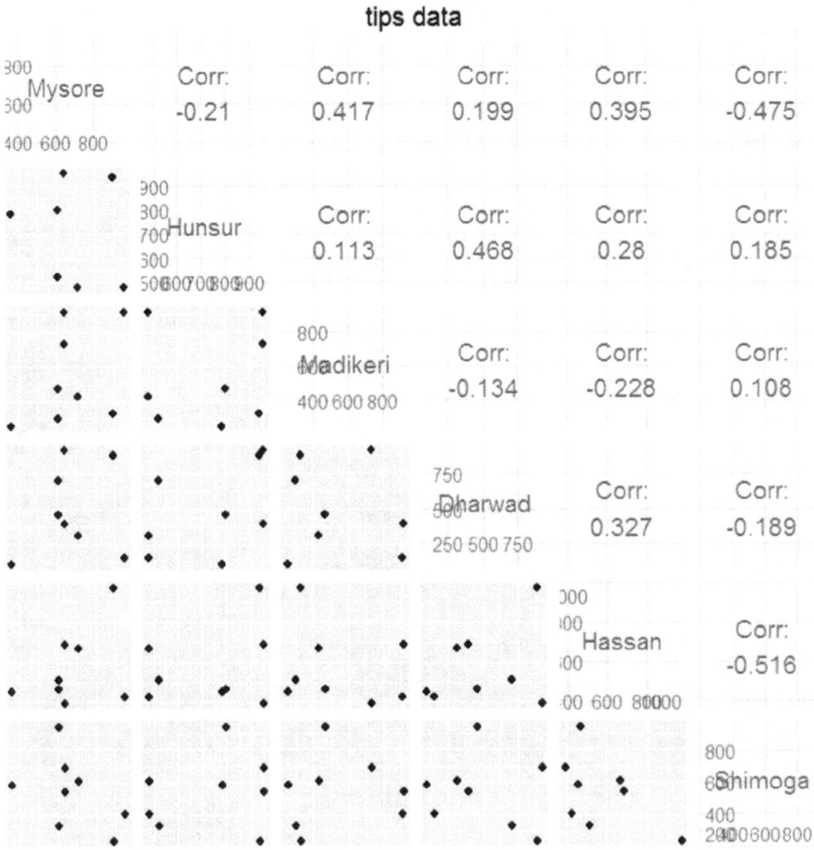

Mosaic Plotslibrary(vcd)

```
aphid=c(210, 1194, 170, 1110, 190, 1406, 730, 1290)
dim(aphid)=c(2, 2, 2)
dimnames(aphid)=
list("Plant Age"=c("Old", "Young"),
"Plant Height"=c("High", "Low"),
"Yield"=c("Yes", "No"))
aphid
library(vcd)
mosaic(aphid, shade=TRUE, legend=TRUE)
```

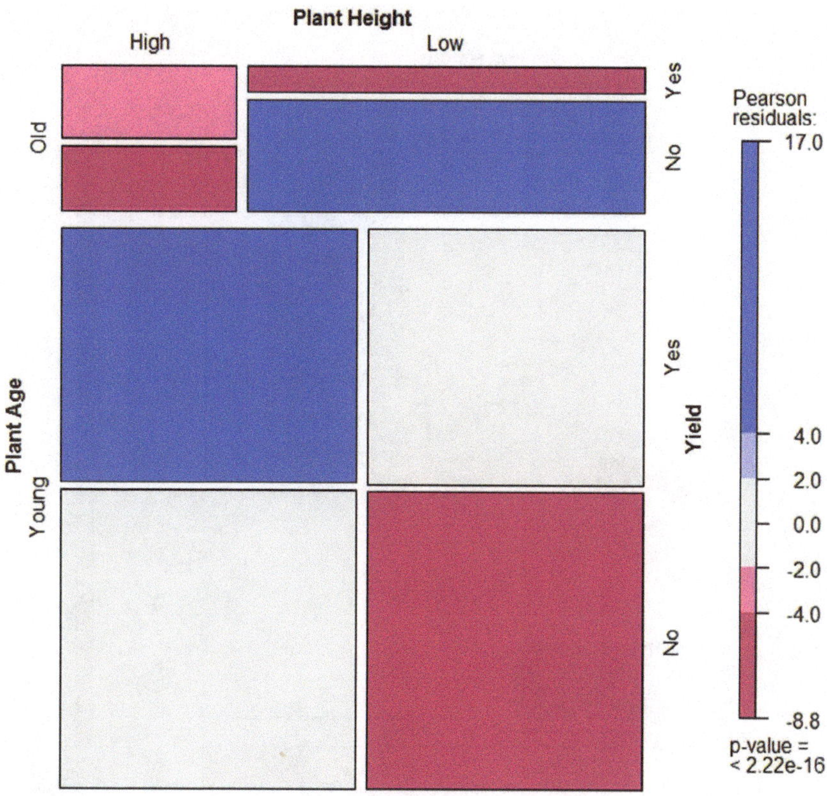

```
aphid = c(210, 1194, 170, 1110, 190, 1406, 730, 1290)
dim(aphid) = c(2, 2, 2)
dimnames(aphid) =
list("Plant Age" = c("Old", "Young"),
"Plant Height" = c("High", "Low"),
"Yield" = c("Yes", "No"))
aphid
library(vcd)
assoc(aphid, shade=TRUE, legend=TRUE)
```

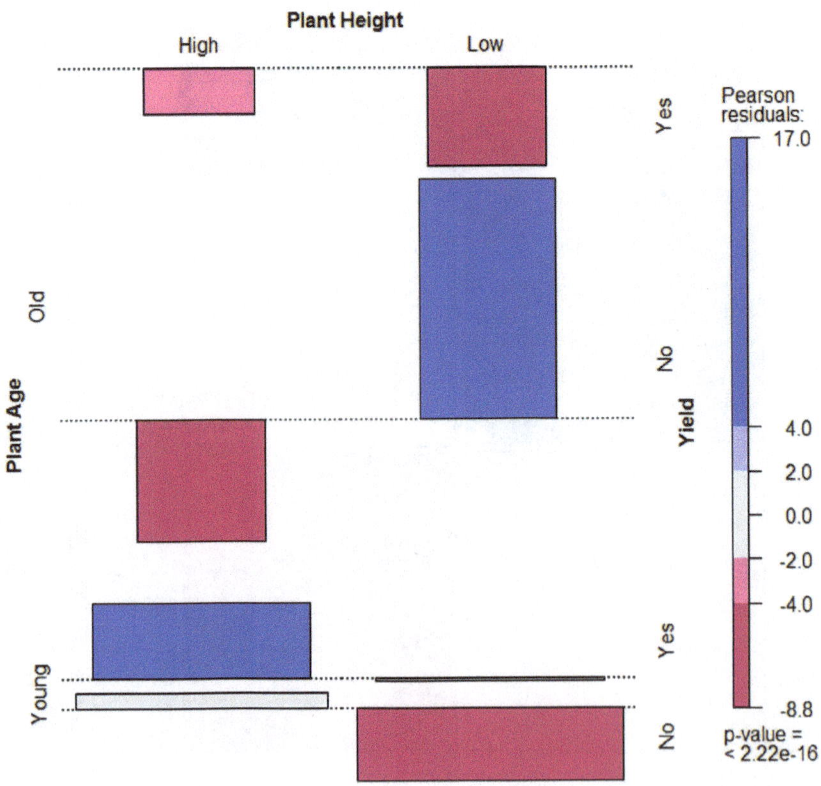

```
library(plotrix)
slices <- c(20, 15, 5, 25, 10)
lbls <- c("Mysore", "Hunsur", "Madikeri", "Dharwad", "Hassan")
pie3D(slices,labels=lbls,explode=0.1,
main="Pie Chart of Cities")
```

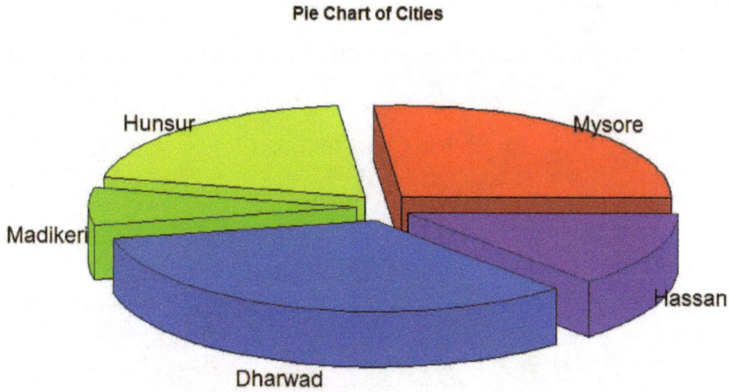


```
# Simple Pie Chart
slices <- c(20, 15, 5, 25, 10)
lbls <- c("Mysore", "Hunsur", "Madikeri", "Dharwad", "Hassan")
pie[slices, labels= lbls, col = c("red", "yellow", "green", "violet","orange", "blue",
title (main = "Pie Chart of Cities"))]
```

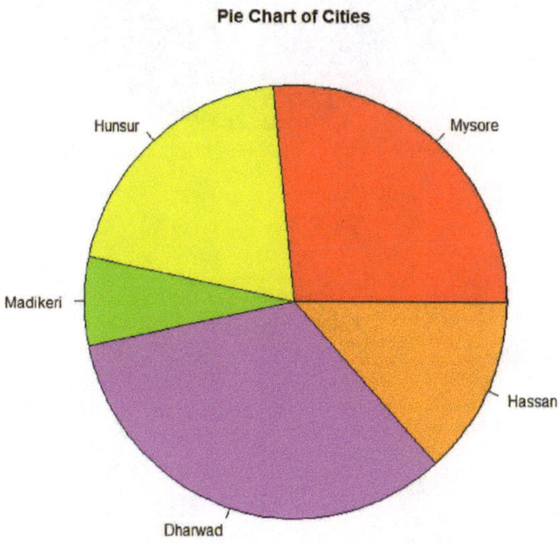

```
# Pie Chart with Percentages
slices <- c(20, 15, 5, 25, 10)
lbls <- c("Mysore", "Hunsur", "Madikeri", "Dharwad", "Hassan")
pct <- round(slices/sum(slices)*100)
lbls <- paste(lbls, pct) # add percents to labels
lbls <- paste(lbls,"%",sep="") # ad % to labels
pie(slices,labels = lbls, col=rainbow(length(lbls)),main= "Pie Chart of Cities")
```

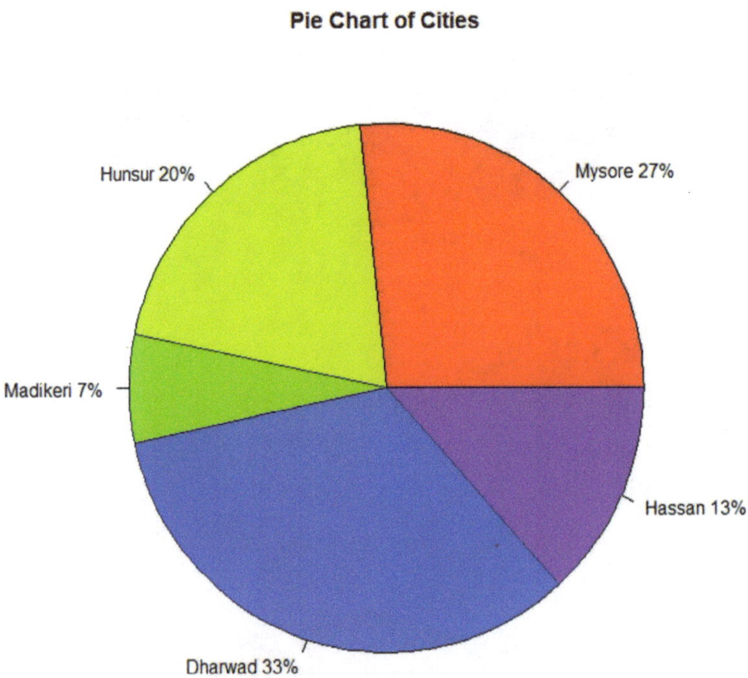

```
# Pie Chart with Percentages
slices <- c(20, 15, 5, 25, 10)
lbls <- c("Mysore", "Hunsur", "Madikeri", "Dharwad", "Hassan")
pct <- round(slices/sum(slices)*100)
lbls <- paste(lbls, pct) # add percents to labels
lbls <- paste(lbls, "%", "sep=") # ad % to labels
pie(slices,labels = lbls, col=rainbow(length(lbls)),main="Pie Chart of Cities")
# Create a legend at the right
legend("topleft", c("Mysore", "Hunsur", "Madikeri", "Dharwad", "Hassan"),
cex=0.8, fill = c("red", "yellow", "green", "violet","orange"))
```

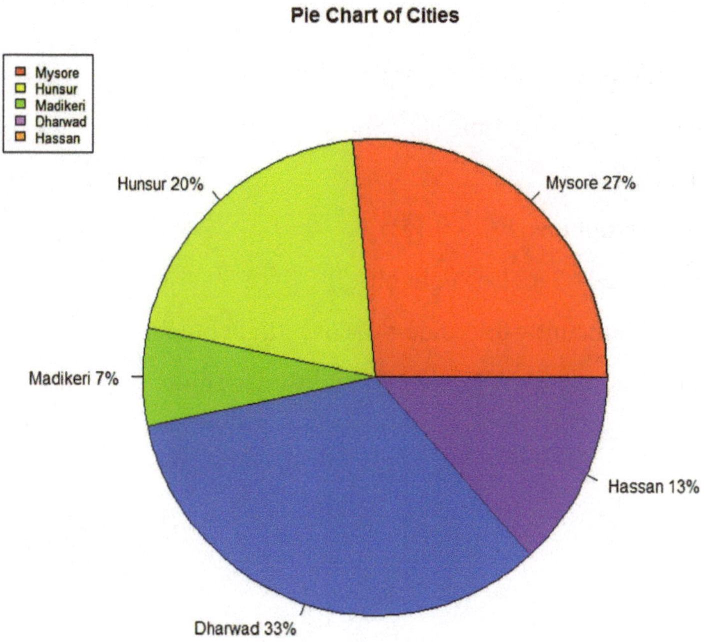

VIOLIN PLOT

```
y7<-(c(2,3,2,3,2,2,1,1,2,1,2,2,0,0,1,0,2,0,1,0,0,0,2, 1,0,0,0,0,0,0,1,1,1, 1,0,0,2,3, 3,
3,3,1,2,1,0,1,0,1,0,1))

y8 <-(c (19,20,20,1 8,0,18, 19,22,22, 18,19, 20,19,17, 16,0,15,17,15,14,20, 21,0,
18, 19,18,17,18,18,16,16,20,21,19,16,18,17,0,18,20,21,18,17,16,0,18,16,17,0))

require(beanplot)

beanplot(y7,y8, ll = 0.02,
      main = "Bean plot", side = "both", xlab="Treatment",
     col = list("purple", c("lightblue", "black")),
      axes=F)

axis(1)

axis(2)

legend("bottomright", fill = c("purple", "lightblue"),
     legend = c("No Treatment", "Sprayed"), box.lty=0)

require(stats)
  # plot(y7(eruptions, bw = 0.15))
  # rug(y7, side = 3, col = "light blue")

rug(jitter(y7, amount = 0.1), side = 4, col = "light blue")

rug(jitter(y7, amount = 0.1), side = 1, col = "dark blue"
```

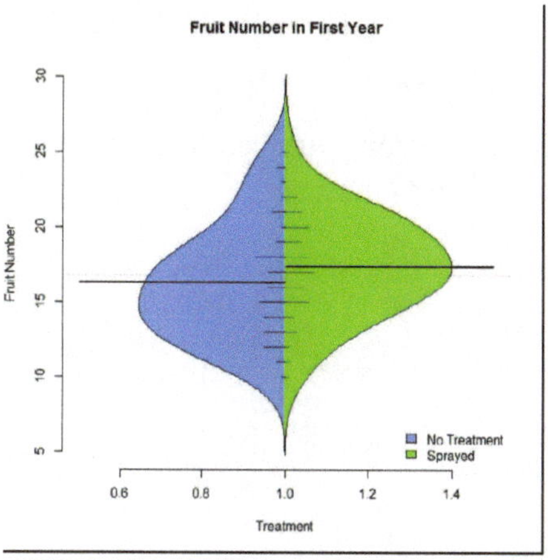
Fruit Number in First Year
Fruit Number
Treatment
No Treatment
Sprayed

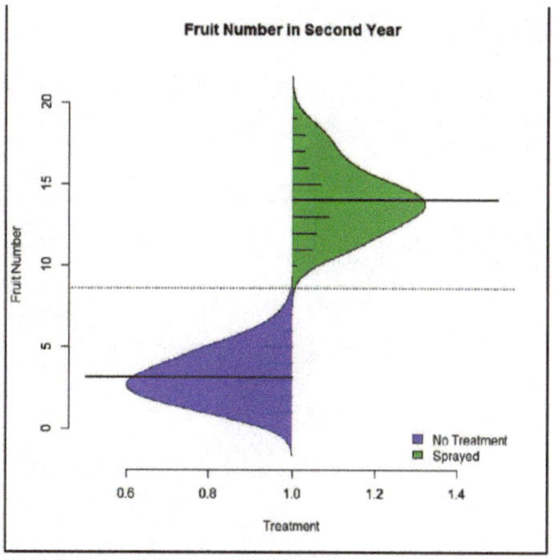
Fruit Number in Second Year
Fruit Number
Treatment
No Treatment
Sprayed

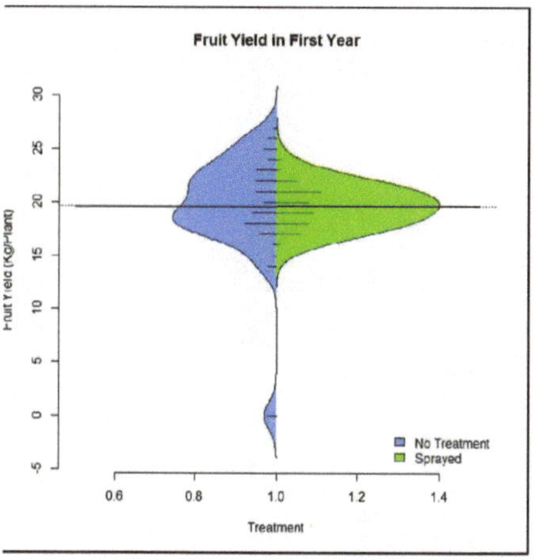
Fruit Yield in First Year
Fruit Yield (Kg/Plant)
Treatment
No Treatment
Sprayed

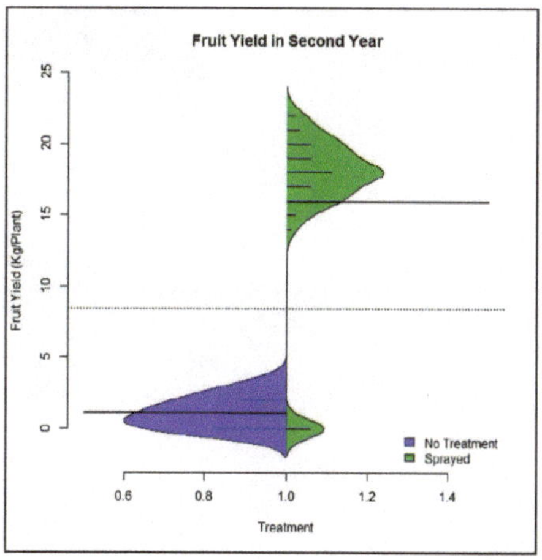
Fruit Yield in Second Year
Fruit Yield (Kg/Plant)
Treatment
No Treatment
Sprayed

y1 <-(c(18,14,12,13,14,15,16,17,18,19,20,21,22,23,24,25,21, 21,24,17, 18,19 ,18,15,16,14,13,14,12,15,12,15,12, 13,14,13,12,11,10,18,15,11,16, 17,18,18,17, 16,15,13))

y2 <-(c(20,21,22,19,18,17,18,15,16,14,13,15,17,14,16,13,17,18,21,22,20,20,20,19, 18,15,17,16,13,12,11,10,15,15,15,18,17,19,16,18,19,16,17,20,21,22,21,20,25,17))

boxplot(y1,y2, notch=TRUE, names = c("Not Sprayed", "Sprayed"),

col=(c("#332288", "#88CCEE")), main = "Fruit Number in First Year", xlab="Effct of Interventional Spray",ylab="Fruit Number")

# Similar for all 4 plots

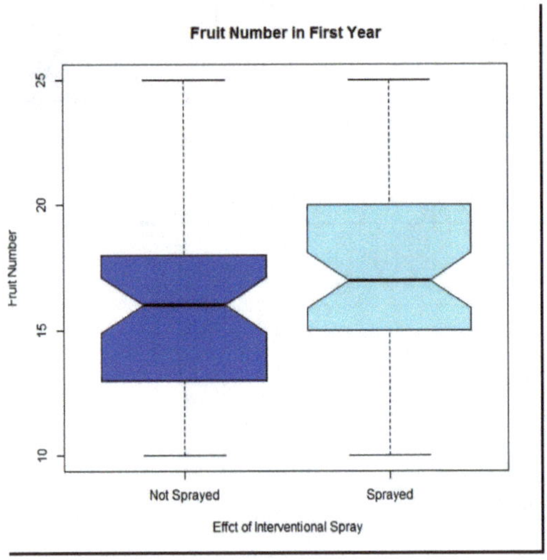


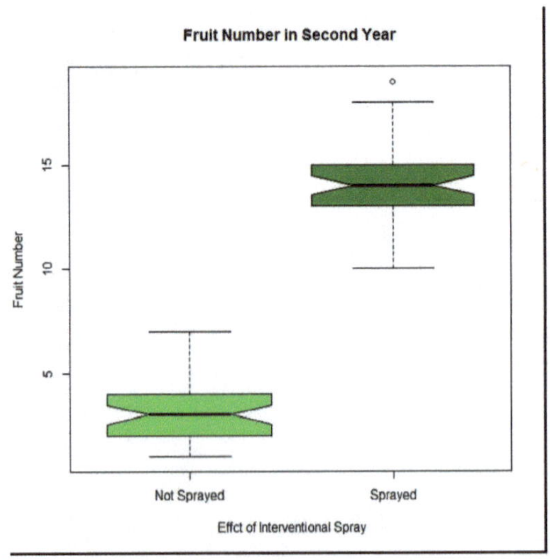

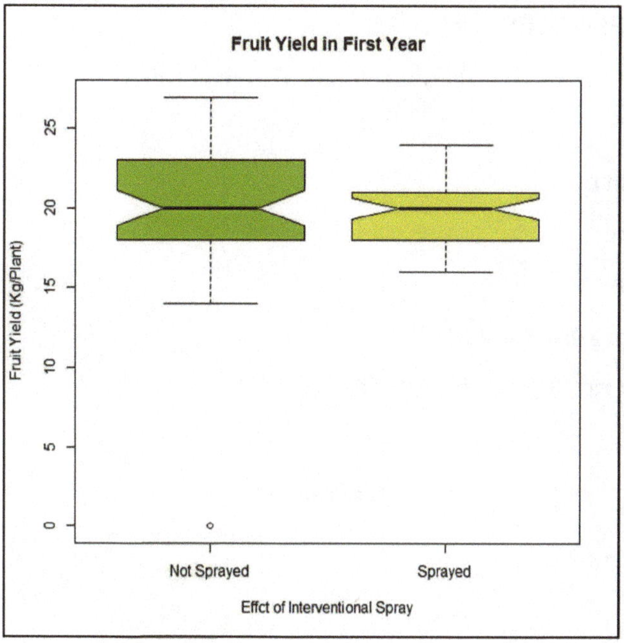
Fruit Yield in First Year
Fruit Yield (Kg/Plant)
0
5
10
15
20
25
Not Sprayed
Sprayed
Effct of Interventional Spray

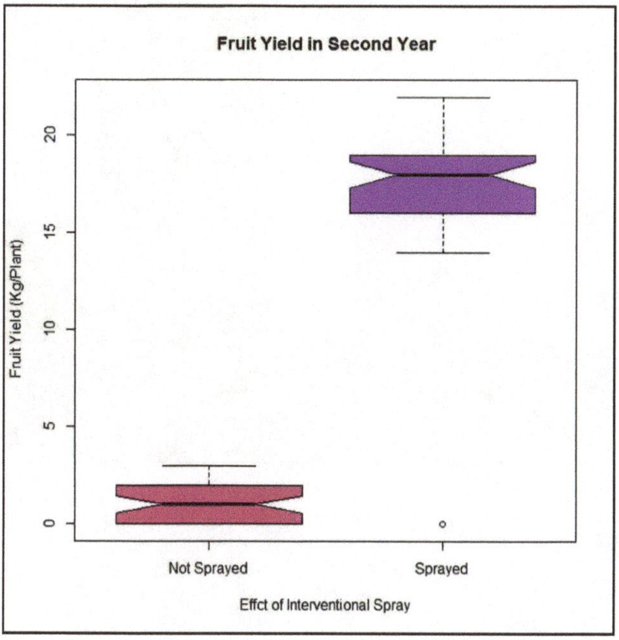
Fruit Yield in Second Year
Fruit Yield (Kg/Plant)
0
5
10
15
20
Not Sprayed
Sprayed
Effct of Interventional Spray

OVERLAPPING HISTOGRAMS

```
my.dat <- read.csv("C:/researchrdata/ 6_col_data.csv")
my.dat
attch(my.dat)
a <- Mysore
b <- Hunsur
hist(a, col=rgb(0,1,0,0.5))
hist(b,col= rgb(1,0,0,0.5),add=T)
```

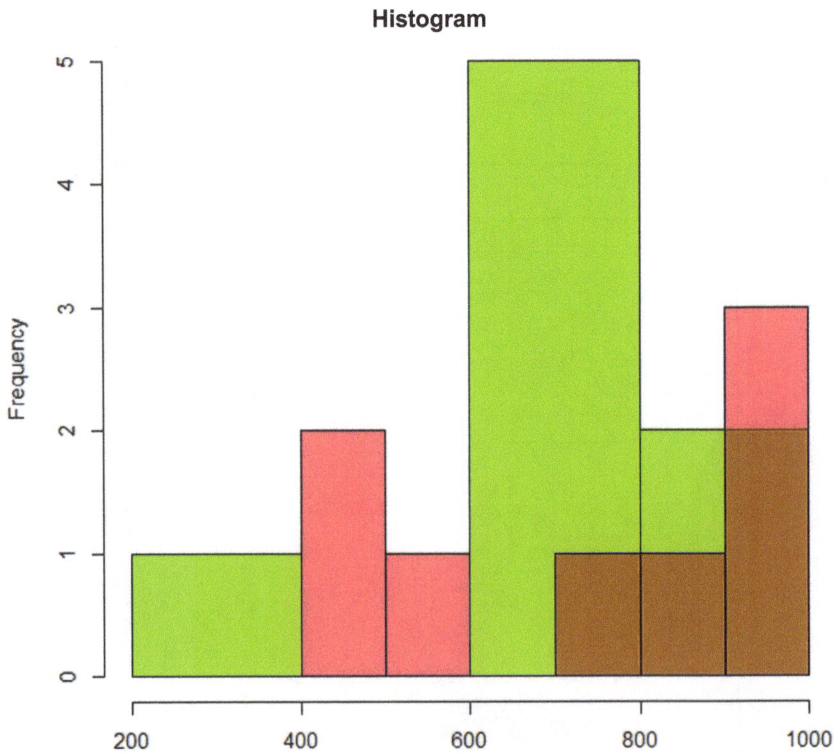

**my.dat <- read.csv("C:/Users/dksamuel_now/Desktop/R-book/data/6_col_data.csv")**

**attach(my.dat)**

**matplot(my.dat, type = c("b"),pch=16,col = 1:6) #plot**

legend("topleft", legend = c("Mysore", "Hunsur", "Madikeri", "Dharwad", "Hassan", "Shimoga"), col=1:6, pch=16) # optional

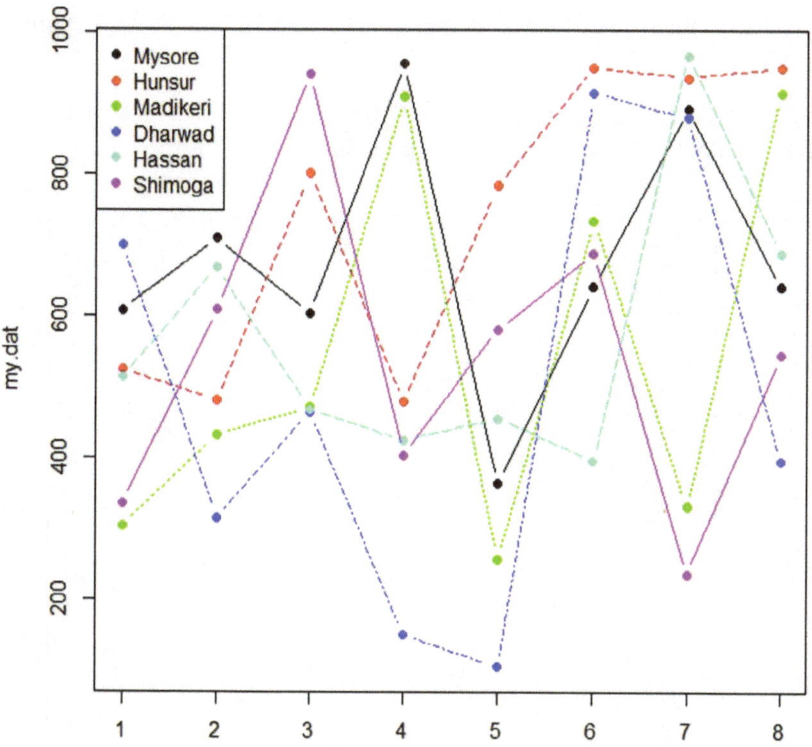

# Similar graph can also be drawn by the following method

x = 1:10 # this value decides the number of points

A = c(15, 36, 54, 60, 68, 71, 73, 75, 78, 78)

B = c(20, 49, 58, 69, 75, 80, 83, 86, 88, 89)

C = c(24, 58, 68, 75, 83, 90, 93, 93, 95, 96)

Performance = data.frame(A,B,C)

matplot(x,Performance, type= "o", pch=c(15,16,17), col=c ("red","green", "blue"))

legend("topleft", legend = c("Mysore", "Hunsur", "Madikeri"), col= c ("red", "green","blue"), pch = c(15,16,17)) # optional

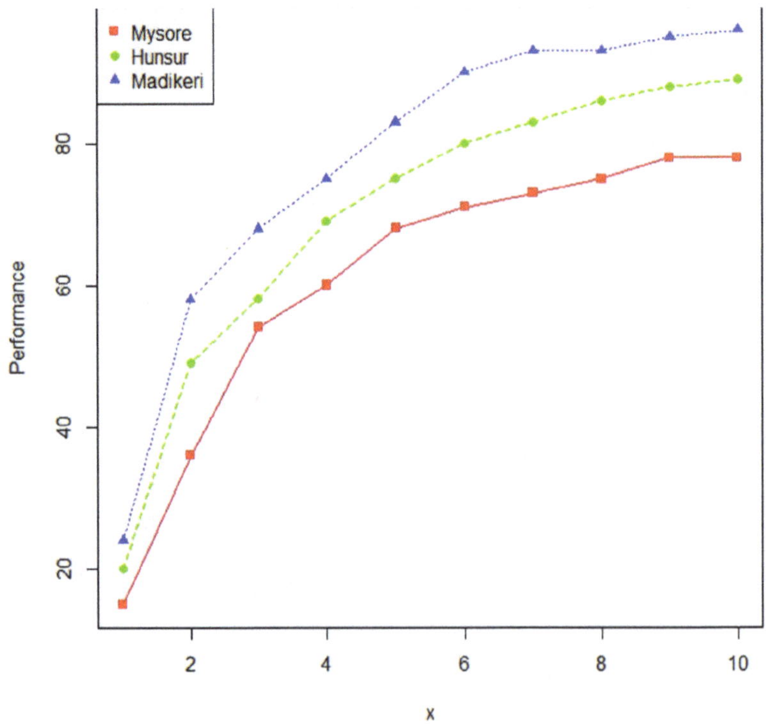

```
# similar plots can be drawn by over-plotting
X <- c(1, 2, 3, 4, 5, 6, 7)
Y1 <- c(2, 4, 5, 7, 12, 14, 16)
Y2 <- c(3, 6, 7, 8, 9, 11, 12)
plot(X, Y1, type= "o", pch = 17, cex=1.2, col= "darkgreen", ylim=c(0,20))
lines(Y2, type= "o", pch=16, lty=2, col= "blue")
title(main= "A PLOT OF TWO VARIABLES", col.main= "red", font.main=2)
```

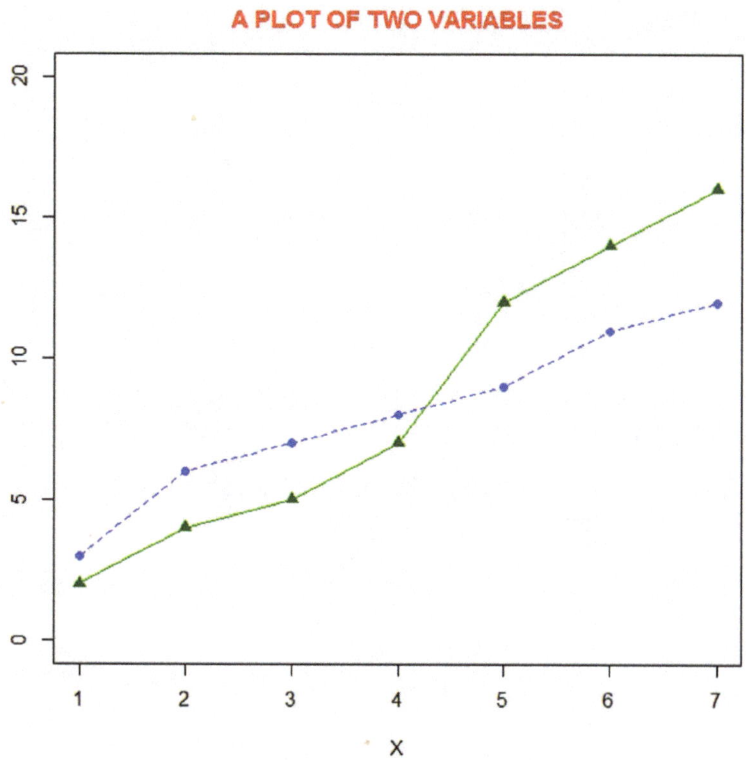

line.x<-c(113,117,235,238,252,263,271,290,300,321,340,354, 369,407,417,427,436,465, 484,494,609,613,622,696,753,763,788,888,974,987,999)

# If values have been pre-sorted from smallest to largest the curve will be smooth

#plot(line.x, type= "l",col= "blue",lty=1,lwd=3, xlab= "Height of plants", ylab= "Place", main= "Lineplot 1")

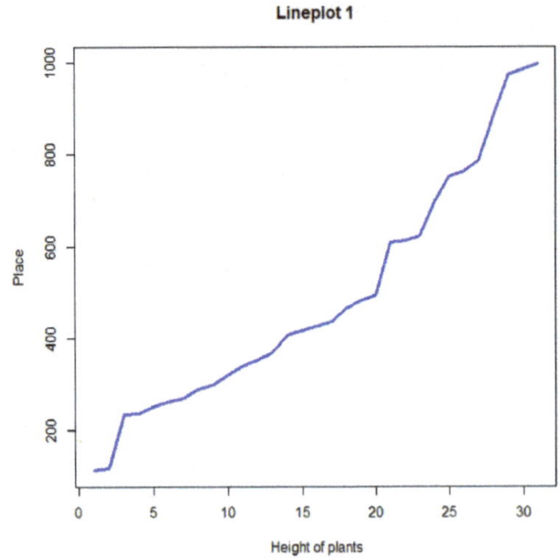


line.x<-c(113,117,235,238,252,263,271,290,300,321, 340,354, 369,407,417,427, 436,465,484,494,609,613,622,696, 753,763,788,888,974,987,999)

plot(line.x, type= "p", col= "green", lty=1,lwd=3, xlab= "Height of plants", ylab= "Place", main= "Lineplot 2")

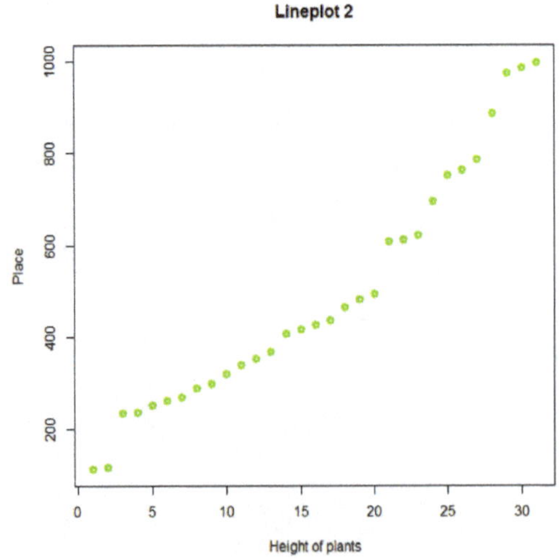

line.x<-c(113,117,235,238,252,263,271,290,300,321,340,354,369,407, 417,427,436,465,484,494,609,613,622,696,753,763,788,888,974,987,999)

plot(line.x, type= “o”, col= “purple”,lty=1,lwd=3, xlab= “Height of plants”, ylab= “Place”, main= “Lineplot 3”)

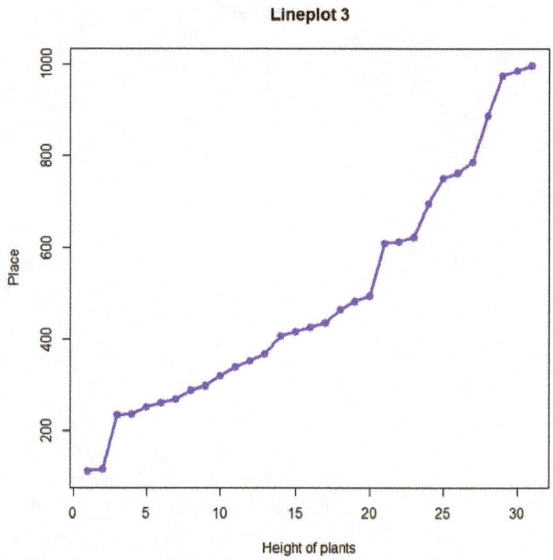


line.x<-c(113,117,235,238,252,263,271,290,300,321,340,354,369, 407,417, 427,436,465,484,494,609,613,622,696,753,763,788,888,974,987,999)

plot(line.x, type= “c”, col= “darkorange”,lty=1,lwd=3, xlab= “Height of plants”, ylab=“Place”, main= “Lineplot 4")

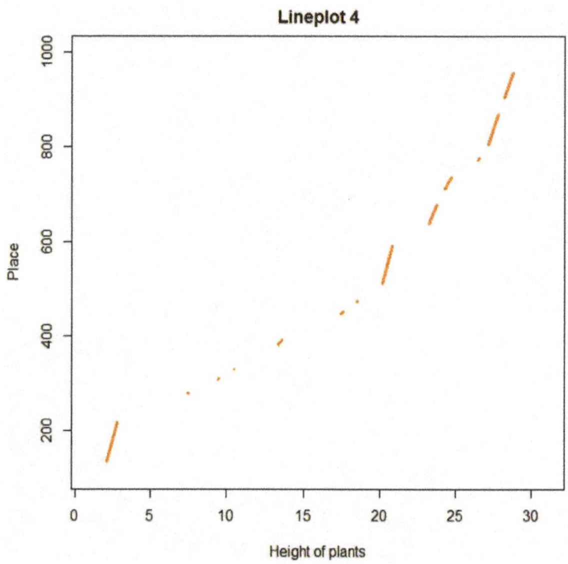

line.x<-c(113,117,235,238,252,263,271,290,300,321,340,354,369,407,417,427, 436,465,484,494,609,613,622,696,753,763,788,888,974,987,999)

plot(line.x, type= "s" , col= "burlywood4", lty=1, lwd=3, xlab= "Height of plants", ylab= "Place", main= "Lineplot 5")

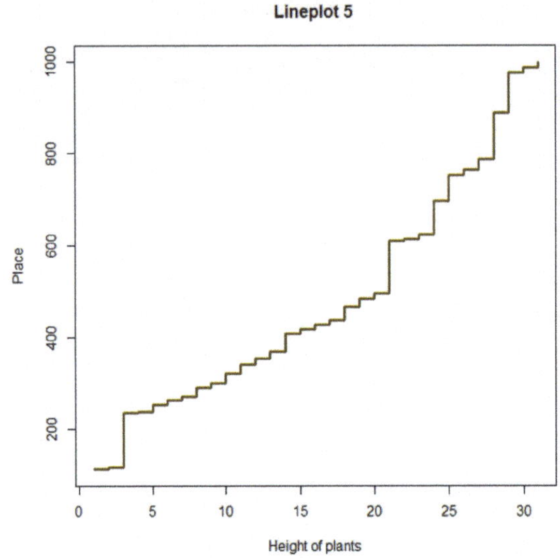


line.x<-c(113,117,235,238,252,263,271,290,300,321,340,354,369, 407,417,427,436,465,484,494,609,613,622,696,753,763,788,888,974,987,999)

plot(line.x, type= "S", col= "darkolivegreen4",lty=1,lwd=3, xlab= "Height of plants", ylab= "Place", main= "Lineplot 6")

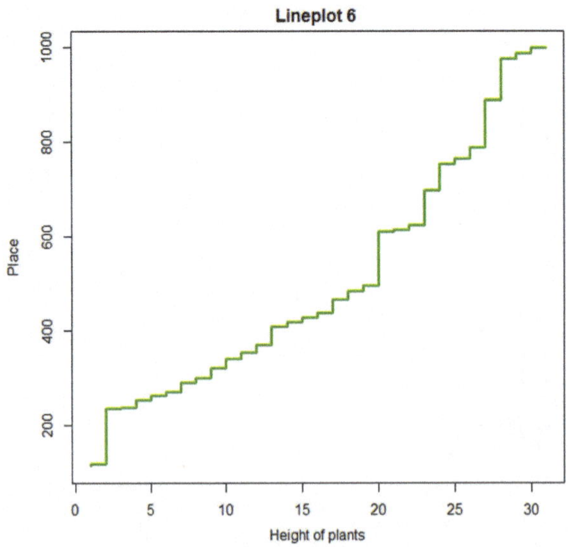

```
line.x<-c(113,117,235,238,252,263,271,290,300,321,340,354,369,407,417,427,
436,465,484,494,609,613,622,696,753,763,788,888,974,987,999)
```

```
plot(line.x, type= “h”, col= “cornflowerblue”,lty=1,lwd=3, xlab= “Height of
plants”, ylab= “Place”, main= “Lineplot 7”)
```

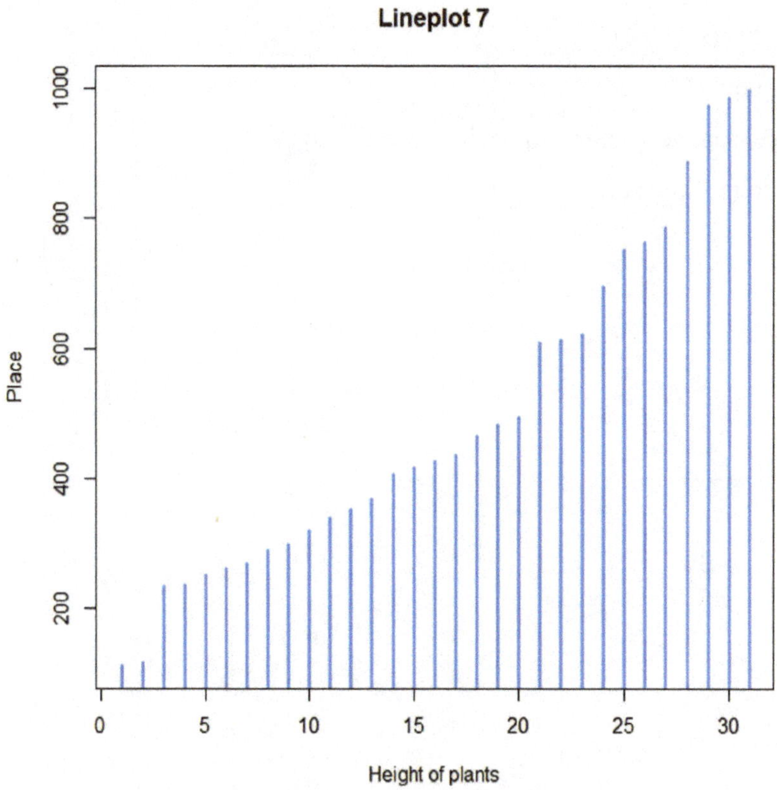

```
library(ggplot2, quietly = TRUE)
# To plot the growth curves for each tree, we need
# to use the group aesthetic:
ggplot(Orange, aes(x = age, y = circumference, group = Tree)) +
geom_line(size = 1) +
geom_point()
# To get separate colors for each line, one way is
last_plot() + aes(color = Tree)
```

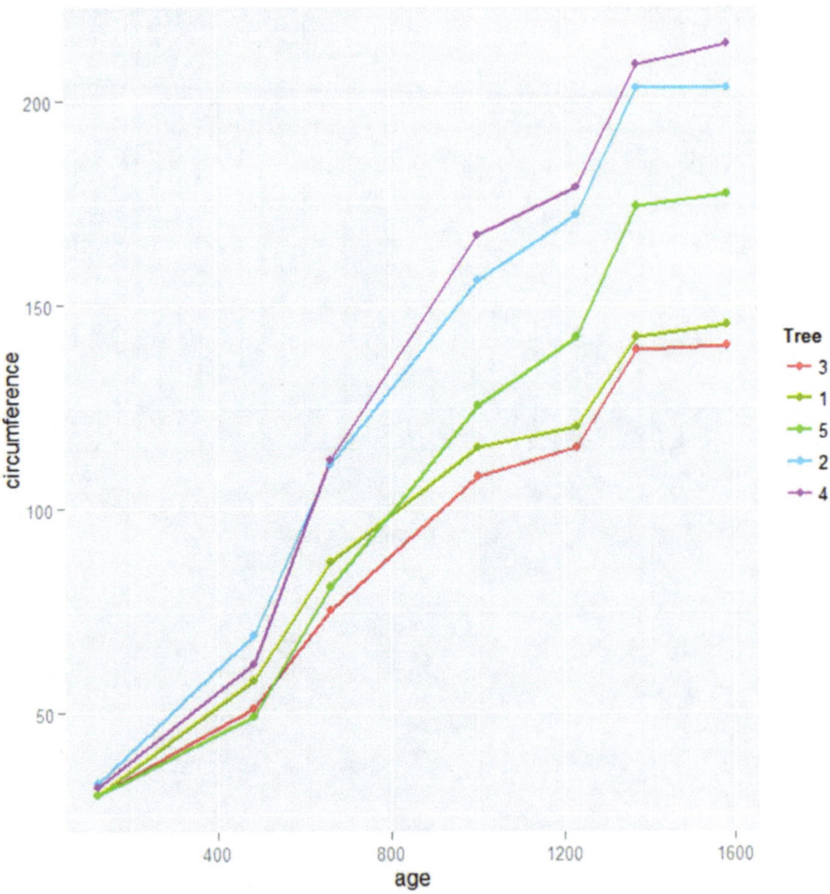

```
t <- read.csv("C:/Users/dksamuel_now/Desktop/R-book/data/orange_11.csv")
qplot(age, circumference, data = t, geom = "line",
colour = Tree,main = "How does orange tree circumference vary with age?")
```

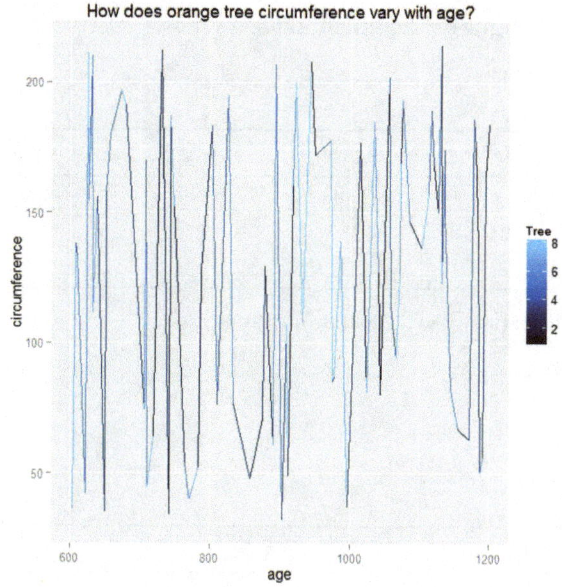


```
qplot(age, circumference, data = t, geom = c("point", "line"), colour = Tree)
```

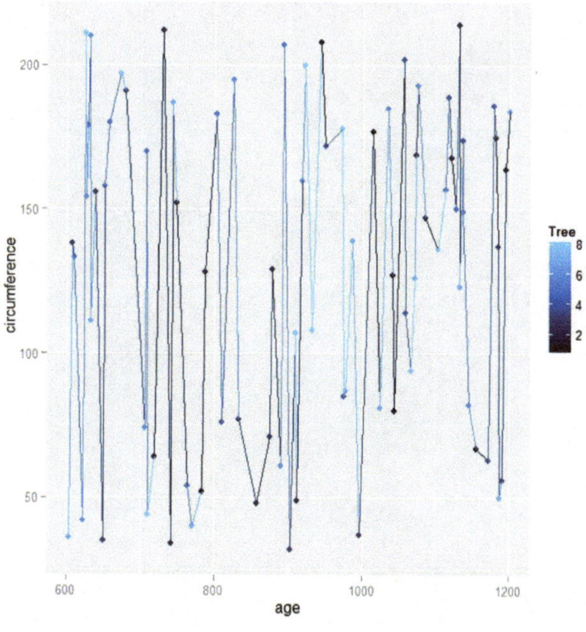

ANOVAS

w<-read.csv(file= "C:/researchrdata/ANOVA_SF.csv", head=TRUE, sep= ",")

# USE HEIGHT ON TREATMENT NOT OTHERWISE

aov.ex1 = aov(Height ~ Treatment, data = w)

summary(aov.ex1)

| | Df | Sum Sq | Mean Sq | F value | Pr(>F) |
|---|---|---|---|---|---|
| Treatment | 4 | 1310847 | 327712 | 1283 | <2e-16 *** |
| Residuals | 25 | 6383 | 255 | | |
| Signif. codes: 0 '***' 0.001 '**' 0.01 '*' 0.05 '.' 0.1 ' ' 1 | | | | | |

print(model.tables(aov.ex1,"means"),digits=3)

Tables of means

| Grand mean |
|---|
| 427 |

| Control | Psuedo_3g | Psuedo_6g | Tricho_3g | Tricho_6g |
|---|---|---|---|---|
| 137 | 638 | 678 | 265 | 417 |

```
boxplot(Height ~ Treatment, data = w)
title(main="Effect of Biocontrol Agents", xlab="Treatments", ylab= "Height in mm")
```

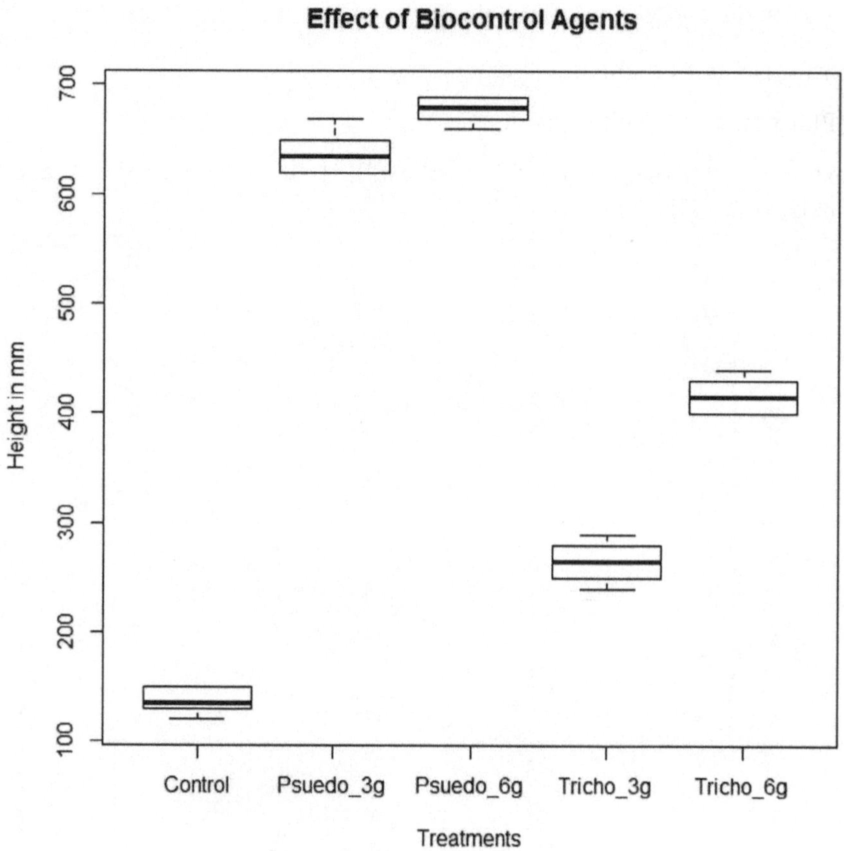

labels <- paste(c("Control" , "Psuedo_3g" , "Psuedo_6g" , "Tricho_3g" , "Tricho_6g"))

boxplot(boxplot(Height ~ Treatment, data = w,col = "lightgray", xaxt = "n", xlab = ",") # x axis with ticks but without labels

axis(1, labels = FALSE)

# Plot x labs at default x position

text(x = seq_along(labels), y = par("usr")[3] - 1, srt = 45, adj = 1,labels = labels, xpd = TRUE)

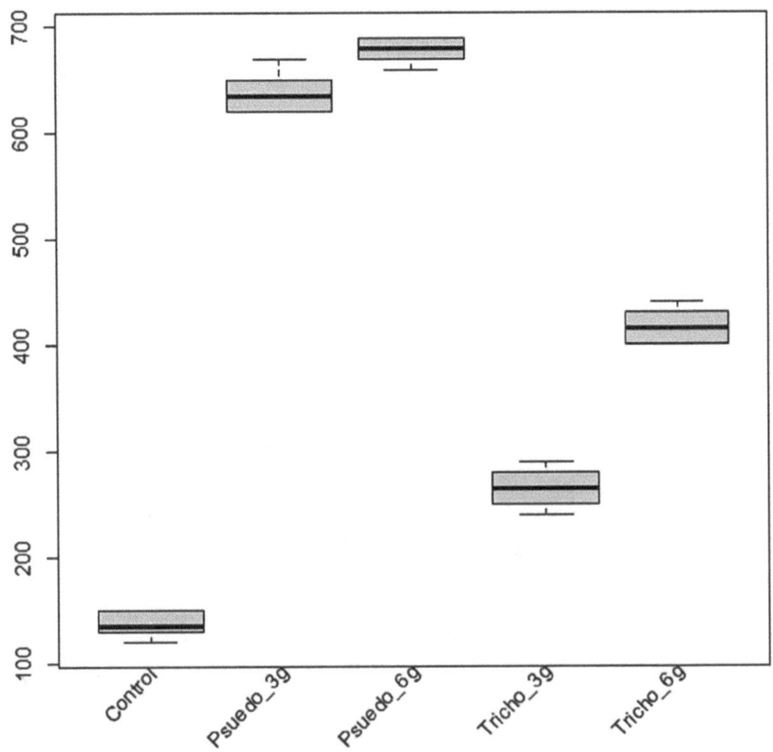

2 way ANOVA

q <-read.csv(file= "C:/researchrdata/ANOVAS223.csv", head=TRUE, sep=",")

aov.2x2 = aov(Height ~ Treatment * Water, data = q)

summary(aov.2x2)

print(model.tables(aov.2x2, "means"),digits=3)

summary(aov.2x2)

| | Df | Sum Sq | Mean Sq | F value | Pr(>F) |
|---|---|---|---|---|---|
| Treatment | 9 | 3945801 | 438422 | 1234 | <2e-16 *** |
| Residuals | 50 | 17757 | 355 | | |
| Signif. codes: 0 '***' 0.001 '**' 0.01 '*' 0.05 '.' 0.1 ' ' 1 | | | | | |

print(model.tables(aov.ex1, "means"), digits=3)

> print(model.tables(aov.2x2, "means"), digits=3)

Tables of means

| Grand mean |
|---|
| 498.1667 |

| Control _I | Control _NI | Psuedo _3g_I | Psuedo _3g_NI | Psuedo _6g_I | Psuedo _6g_NI | Tricho _3g_I | Tricho _3g_NI | Tricho_ 6g_I | Tricho _6g_NI |
|---|---|---|---|---|---|---|---|---|---|
| 137 | 6182 | 638 | 851 | 678 | 904 | 265 | 353 | 417 | 555 |

interaction.plot(Water,Treatment,Height)

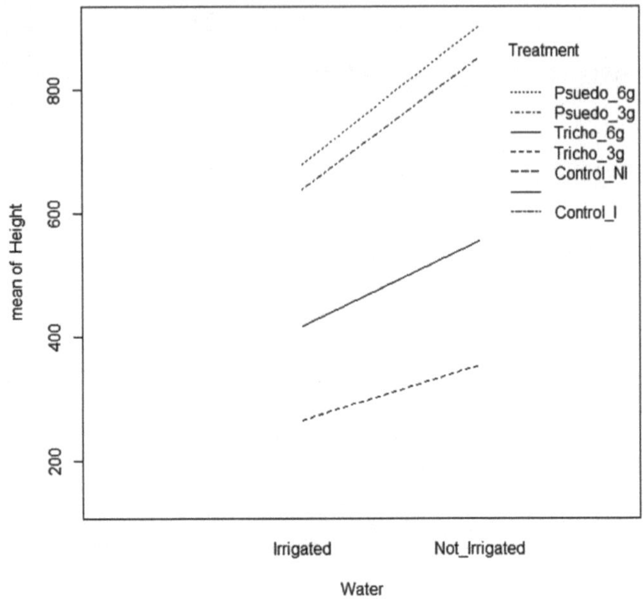


USE WITHOUT LABELS BECAUSE THE SUMMARY GIVES THE LABEL SEQUNCE

boxplot(Height ~ Treatment * Water, data = q, col = "lightgray", xaxt = "n", xlab = ",") # x axis with ticks but without labels

**axis(1, labels = FALSE)**

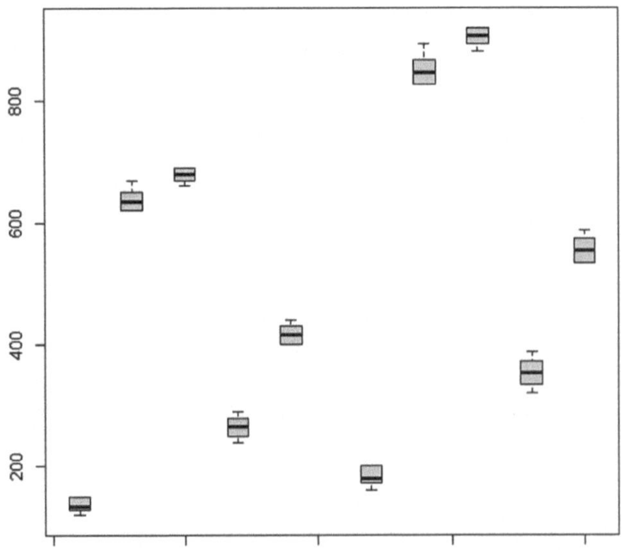

NEWANOVA

```
k <-read.csv(file= "C:/researchrdata/ANOVA_SF.csv", head=TRUE, sep= ",")
attach(k)
par(mfrow=c(1,2))
plot(Yield ~ Irrigation + Spray, data = k)
```

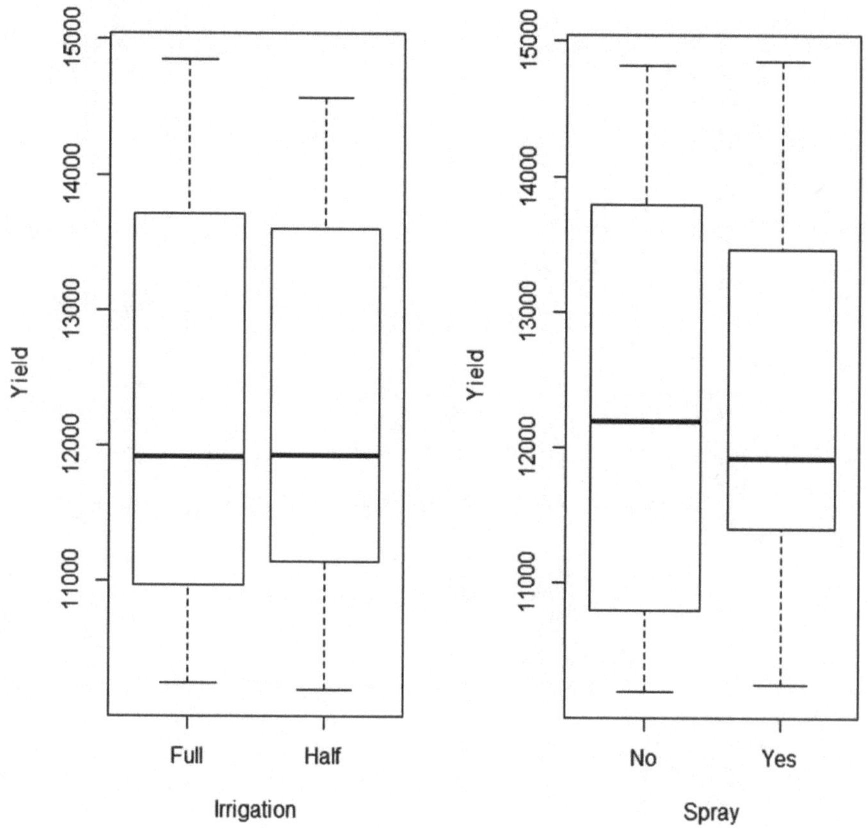

```
interaction.plot (Spray, Irrigation, Yield)
```

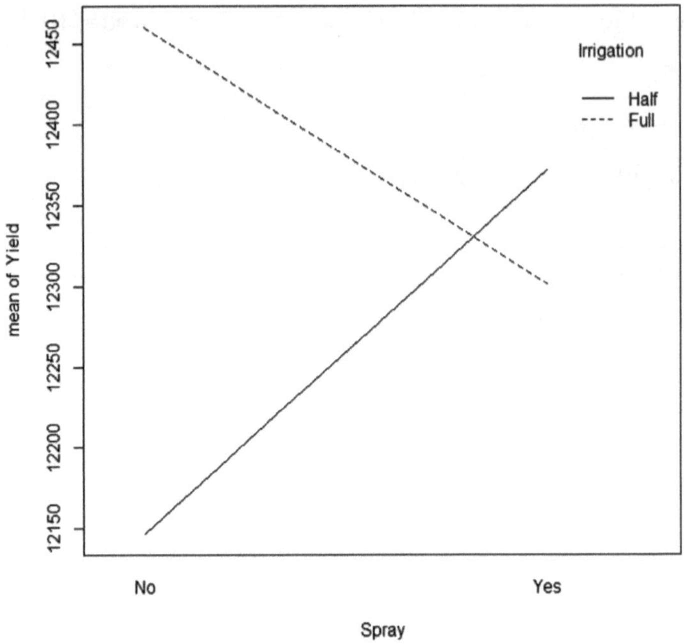


```
interaction.plot (Irrigation, Spray, Yield)
```

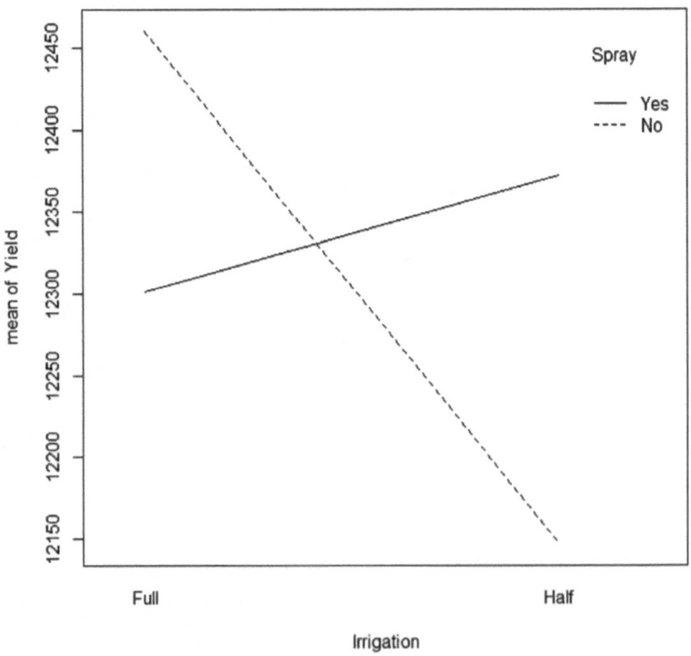

k.model = lm(Yield ~ Spray + Irrigation + Spray*Irrigation)

anova(k.model)

Analysis of Variance Table: Response: Yield

| | Df | Sum Sq | Mean Sq | F value | Pr(>F) |
|---|---|---|---|---|---|
| Spray | 1 | 10791 | 10791 | 0.004 | 0.9449 |
| Irrrigation | 1 | 147987 | 147987 | 0.0665 | 0.7979 |
| Spray:Irrigation | 1 | 370755 | 370755 | 0.1667 | 0.6855 |
| Residuals | 36 | 80064467 | 2224013 | | |

l.model = lm(Yield ~ Irrigation + Spray + Irrigation*Spray)

anova(l.model)

Analysis of Variance Table: Response: Yield

| | Df | Sum Sq | Mean Sq | F value | Pr(>F) |
|---|---|---|---|---|---|
| Irrigation | 1 | 147987 | 147987 | 0.0065 | 0.7979 |
| Spray | 1 | 10791 | 10791 | 10791 | 10791 |
| Irrigation: Spray | 1 | 370755 | 370755 | 0.1667 | 0.6855 |
| Residuals | 36 | 80064467 | 2224013 | | |

qqnorm(k.model$res)

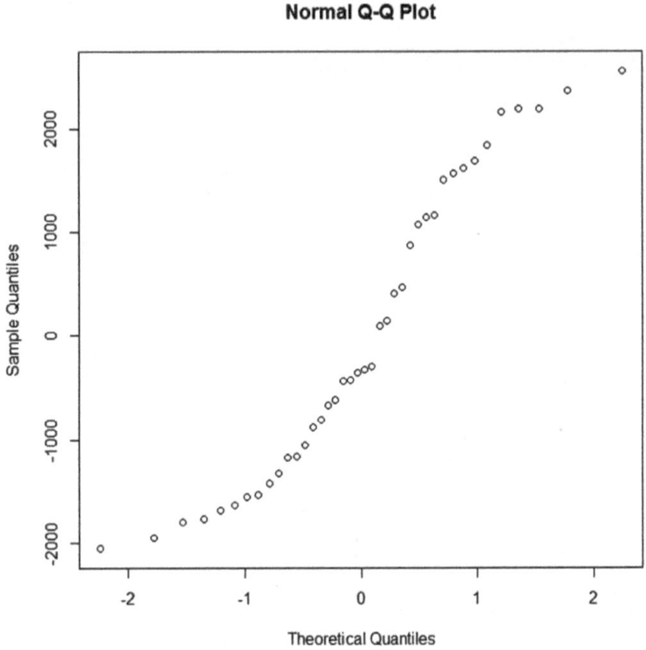


qqnorm(l.model$res)

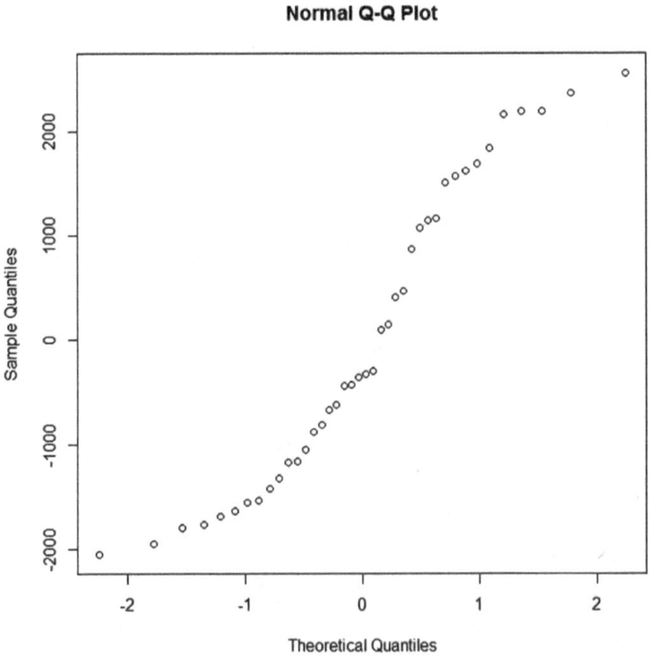

plot(k.model$fitted, k.model$res, xlab="Fitted", ylab="Residuals")

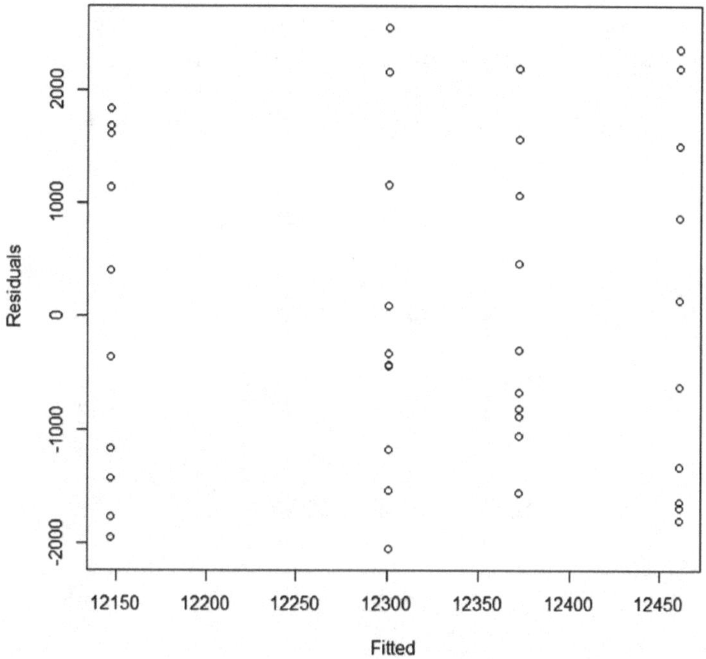


plot(l.model$fitted, l.model$res, xlab= "Fitted", ylab= "Residuals")

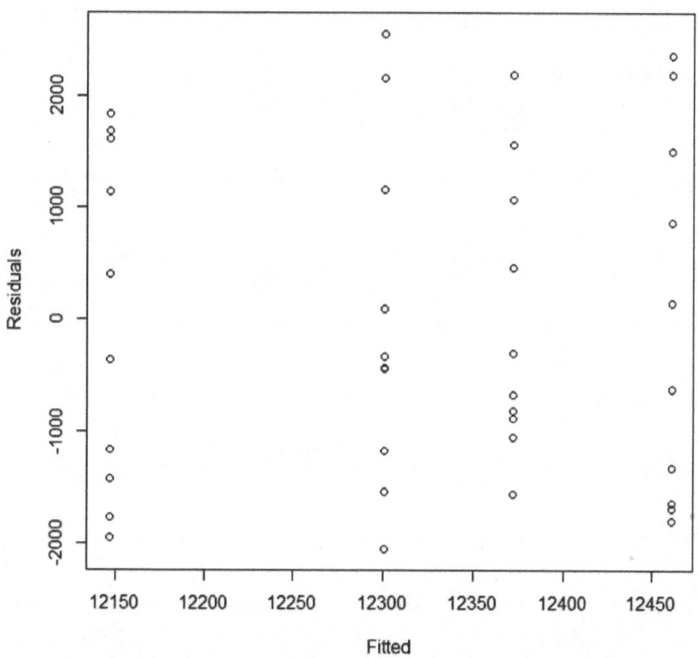

Plotting Symbols

Use the **pch=** option to specify symbols to use when plotting points. For symbols 21 through 25, specify border color (col=) and fill color (bg=).

Lines

You can change lines using the following options. This is particularly useful for reference lines, axes, and fit lines.

| option | description |
|---|---|
| **lty** | line type. see the chart below. |
| **lwd** | line width relative to the default (default=1). 2 is twice as wide. |

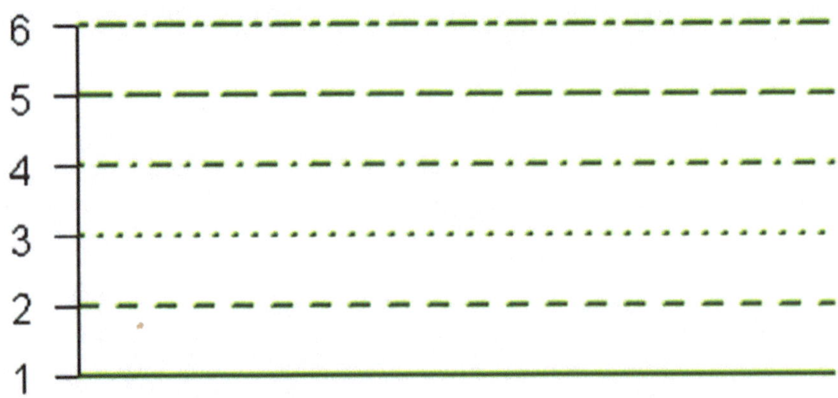

For example **col=1**, **col= "white"**, and **col= "#FFFFFF"** are equivalent.

The following chart was produced with code developed by Earl F. Glynn. See his **Color Chart** for all the details you would ever need about using colors in **R**.

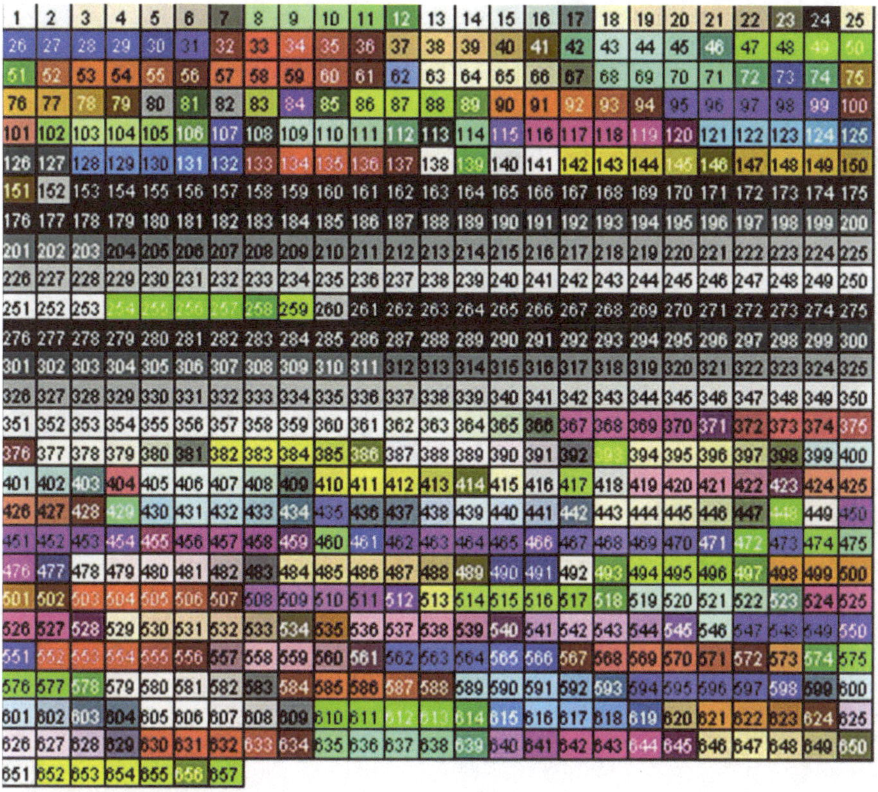


You can also create a vector of *n* contiguous colors using the functions **rainbow(*n*)**, **heat.colors(*n*)**,**terrain.colors(*n*)**, **topo.colors(*n*)**, and **cm.colors(*n*)**.

**colors()** returns all available color names.

http://html-color-codes.com/

ADD Color Chart:

http://research.stowers-institute.org/efg/R/Color/Chart/ColorChart.pdf

COLOR BREWER

———

library(RColorBrewer)

display.brewer.all()

display.brewer.pal(n = 6, name = “Spectral”)

display.brewer.pal(n = 6, name = “BrBG”)

color <-brewer.pal(n = 6, name = “BrBG”)# will show the colors

color# willgive the names of the colors like [1] “#8C510A” “#D8B365” “#F6E8C3” “#C7EAE5” “#5AB4AC” “#01665E”

———Very nice example

data(VADeaths)

par(mfrow=c(2,3))

hist(VADeaths,col=brewer.pal(3,“Set3"),main= “Set3 3 colors”)

hist(VADeaths,col=brewer.pal(3, “Set2"),main= “Set2 3 colors”)

hist(VADeaths,col=brewer.pal(3, “Set1"),main= “Set1 3 colors”)

hist(VADeaths,col=brewer.pal(8, “Set3"),main= “Set3 8 colors”)

hist(VADeaths,col=brewer.pal(8, “Greys”),main= “Greys 8 colors”)

hist(VADeaths,col=brewer.pal(8, “Greens”),main= “Greens 8 colors”)

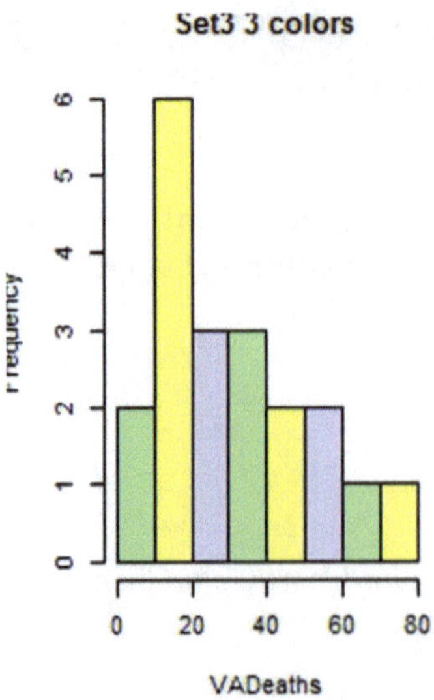


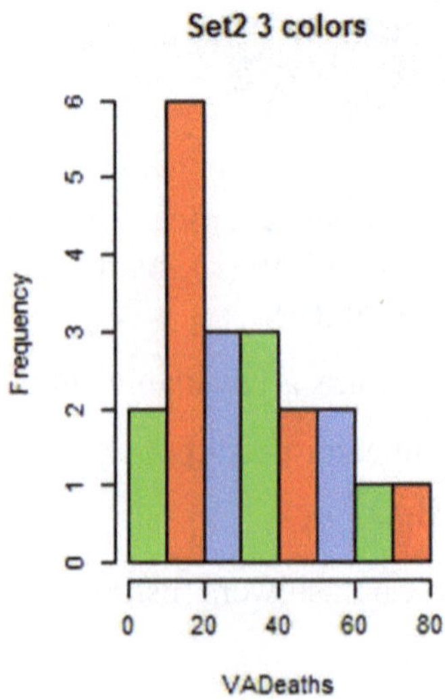

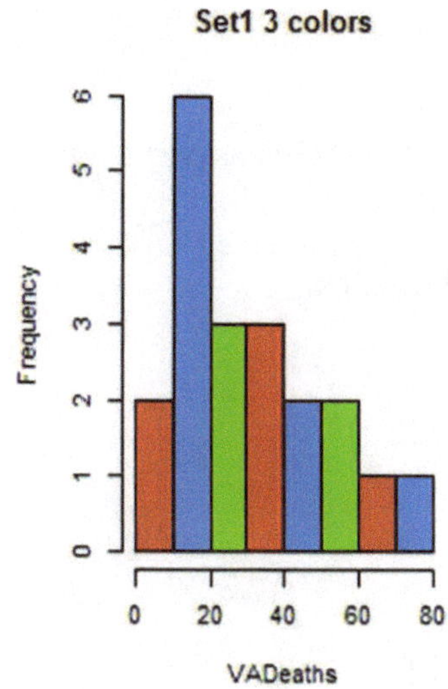


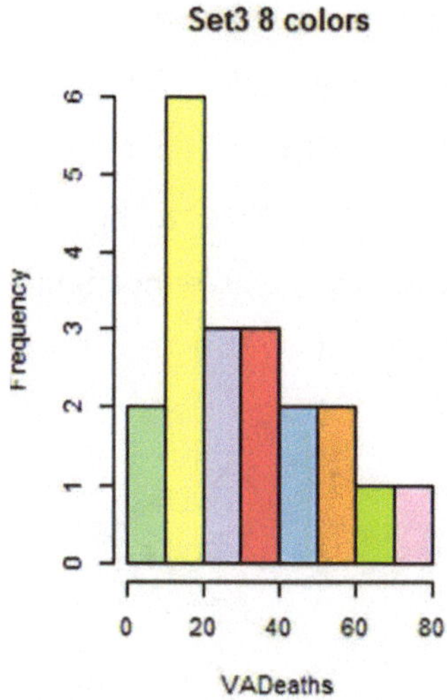


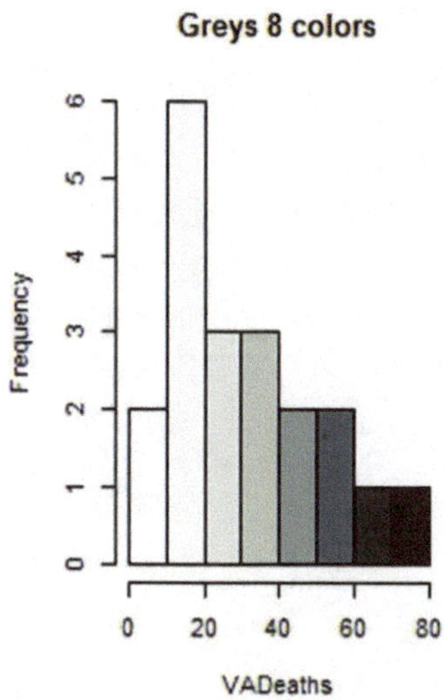


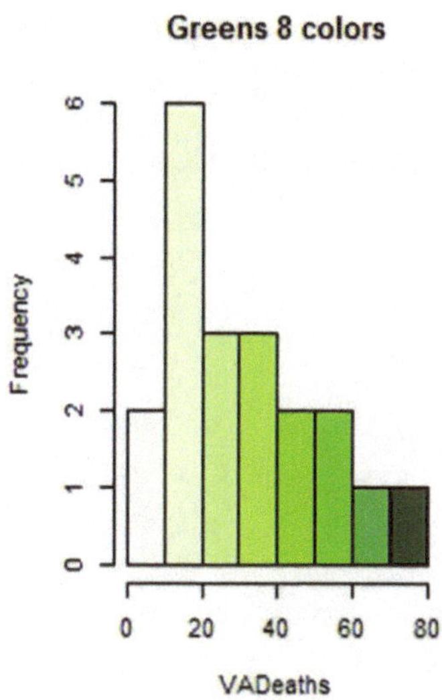


http://decisionstats.com/2012/04/08/color-palettes-in-r-using-rcolorbrewer-rstats/

```
## create a sequential palette for reuse and show it
mypalette<-brewer.pal(7, "Greens")
image(1:7,1,as.matrix(1:7),col=mypalette,xlab= "Greens (sequential)",
ylab= "" xaxt= "n",yaxt= "n",bty= "n")
> dat <- read.table(textConnection("ID Day Dose
+     61552     1   .9
+     71552     7   .8
+     81552    14   3.5
+     91663     1   2.2
+     71663     7   3.7
+     51663    18   5.4"), header=TRUE)
> require(ggplot2)
> p <- ggplot(dat, aes(factor(Day), factor(ID)))
> p + geom_dotplot(binaxis = "y", stackdir = "center", binpositions= "all") +
+ geom_point(aes(size=Dose))
```

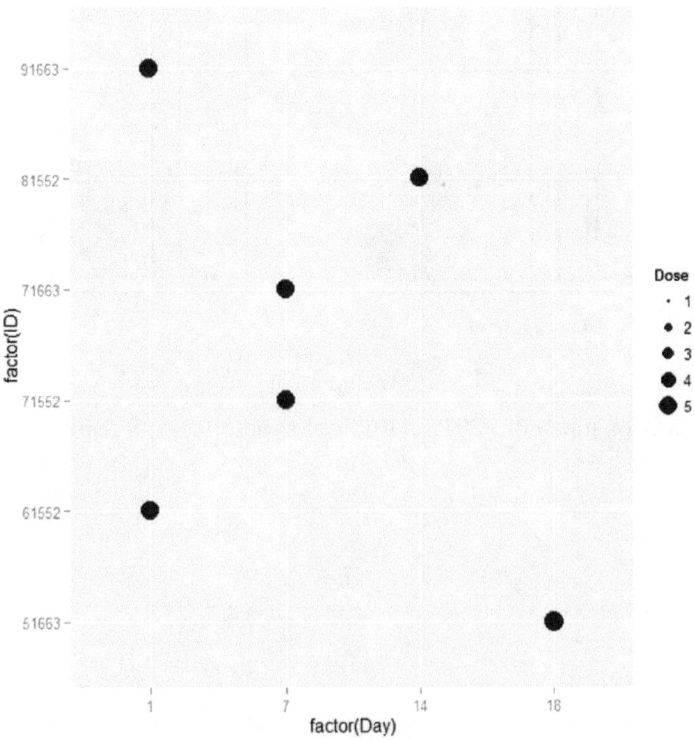

## The following dataset is the time required for a

## group of farmers to complete a field operation

farm.time=c(215, 312, 139, 195, 318, 217, 168, 318, 228, 213, 311, 414, 213, 317, 517)

stripchart(farm.time,

method= "stack",

pch=19,

main= "Time Taken to Complete Farm work",

col= "Blue",

offset=0.5,

xlab= "Time (sec)")

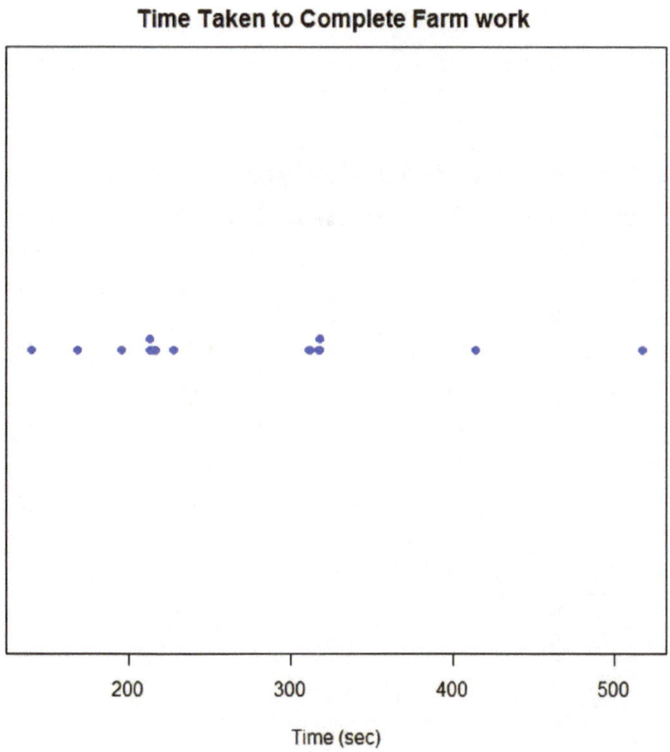

```
q <-read.csv(file= "C:/Users/dksamuel_now/Desktop/R-book/data/dotchart.csv", head=TRUE, sep=",")
dotchart((t(q)),pch = 16, col = "blue")
```

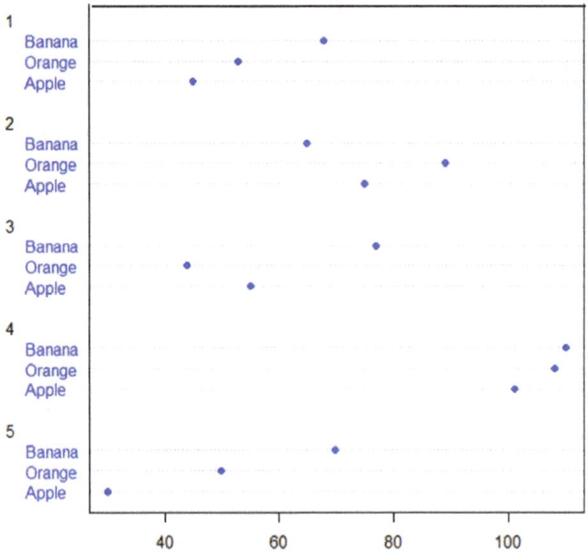


```
dotchart(t(q), color=c("red", "blue", "darkgreen"),
main= "Dotchart for Fruits", cex=0.8, pch= 16)
```

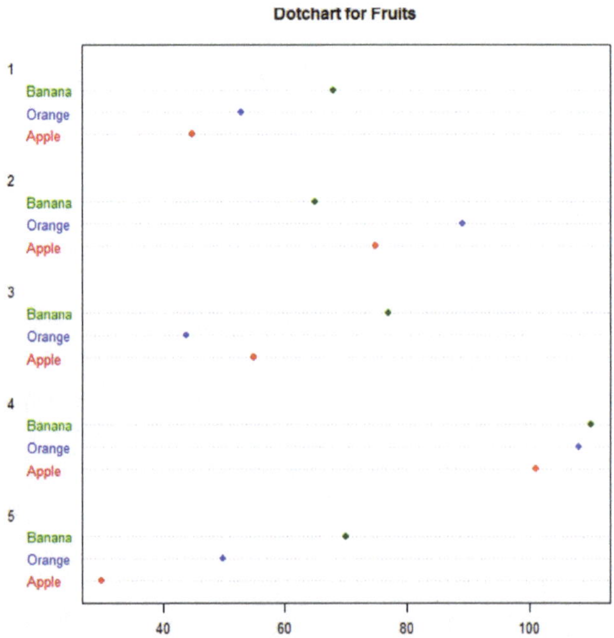

**AGRICOLAE (it is a free R package, you can read about it in**

| | virus | yield |
|---|---|---|
| 1 | Kolar | 3289 |
| 2 | Kolar | 4823 |
| 3 | Kolar | 3232 |
| 4 | Gulburga | 5364 |
| 5 | Gulburga | 6351 |
| 6 | Gulburga | 5323 |
| 7 | Mysore | 7421 |
| 8 | Mysore | 6839 |
| 9 | Mysore | 6983 |
| 10 | Mandya | 6785 |
| 11 | Mandya | 6200 |
| 12 | Mandya | 6803 |

```
model<-aov(yield~virus, data=my.dat)
cv.model(model)
[1] 10.09143
mean(yield)
[1] 5784.417
df<-df.residual(model)
MSerror<-deviance(model)/df
# comparison <- LSD.test(yield,virus,df,MSerror)
LSD.test(model, "virus",console=TRUE)
Study: model ~ "virus"
LSD t Test for yield
Mean Square Error: 340740.9
```

virus, means and individual (95 %) CI

| | yield | std | r | LCL | UCL | Min | Max |
|---|---|---|---|---|---|---|---|
| Gulburga | 5679.333 | 582.0415 | 3 | 4902.171 | 6456.495 | 5323 | 6351 |
| Kolar | 3781.333 | 902.5599 | 3 | 3004.171 | 4558.495 | 3232 | 4823 |
| Mandya | 6596.000 | 343.0641 | 3 | 5818.838 | 7373.162 | 6200 | 6803 |
| Mysore | 7081.000 | 303.1237 | 3 | 6303.838 | 7858.162 | 6839 | 7421 |

alpha: 0.05 ; Df Error: 8

Critical Value of t: 2.306004

Least Significant Difference 1099.073

Means with the same letter are not significantly different

| Groups | Treatments | means |
|---|---|---|
| a | Mysore | 7081 |
| ab | Mandya | 6596 |
| b | Gulburga | 5679 |

# comparison <- LSD.test(yield, virus,df, MSerror, group=F)

outLSD <-LSD.test(model, “virus”, group=F,console=TRUE)

Study: model ~ “virus”

LSD t Test for yield

Mean Square Error: 340740.9

virus, means and individual ( 95 %) CI

| | Yield | STD | R | LCL | UCL | Min | Max |
|---|---|---|---|---|---|---|---|
| Gulburga | 5679.333 | 582.0415 | 3 | 4902.171 | 6456.495 | 5323 | 6351 |
| Kolar | 3781.333 | 902.5599 | 3 | 3004.171 | 4558.495 | 3232 | 4823 |
| Mandya | 6596.000 | 343.0641 | 3 | 5818.838 | 7373.162 | 6200 | 6803 |
| Mysore | 7081.000 | 303.1237 | 3 | 6303.838 | 7858.162 | 6839 | 7421 |

Critical Value of t: 2.306004

Comparison between treatments means

| | Difference | Pvalue | Sig | LCL | UCL |
|---|---|---|---|---|---|
| Gulburga - Kolar | 1898.0000 | 0.0040488757 | ** | 798.9269 | 2997.0731 |
| Gulburga - Mandya | -916.6667 | 0.0906542516 | | -2015.7398 | 182.4065 |
| Gulburga - Mysore | -1401.6667 | 0.0186842946 | * | -2500.7398 | -302.5935 |
| Kolar - Mandya | -2814.6667 | 0.0003594899 | *** | -3913.7398 | -1715.5935 |
| Kolar - Mysore | -3299.6667 | 0.0001216575 | *** | -4398.7398 | -2200.5935 |
| Mandya - Mysore | -485.0000 | 0.3386540923 | | -1584.0731 | 614.0731 |

**print(outLSD)**

**$statistics**

| Mean | CV | MSerror |
|---|---|---|
| 5784.417 | 10.09 | 340740.9 |

**$parameters**

| DF | NTR | t.value |
|---|---|---|
| 8 | 4 | 2.306004 |

**$means**

| | Yield | STD | R | LCL | UCL | Min | Max |
|---|---|---|---|---|---|---|---|
| Gulburga | 5679.333 | 582.0415 | 3 | 4902.171 | 6456.495 | 5323 | 6351 |
| Kolar | 3781.333 | 902.5599 | 3 | 3004.171 | 4558.495 | 3232 | 4823 |
| Mandya | 6596.000 | 343.0641 | 3 | 5818.838 | 7373.162 | 6200 | 6803 |
| Mysore | 7081.000 | 303.1237 | 3 | 6303.838 | 7858.162 | 6839 | 7421 |

**$comparison**


| | Difference | Pvalue | Sig | LCL | UCL |
|---|---|---|---|---|---|
| Gulburga - Kolar | 1898.0000 | 0.0040488757 | ** | 798.9269 | 2997.0731 |
| Gulburga - Mandya | -916.6667 | 0.0906542516 | | -2015.7398 | 182.4065 |
| Gulburga - Mysore | -1401.6667 | 0.0186842946 | * | -2500.7398 | -302.5935 |
| Kolar - Mandya | -2814.6667 | 0.0003594899 | *** | -3913.7398 | -1715.5935 |
| Kolar - Mysore | -3299.6667 | 0.0001216575 | *** | -4398.7398 | -2200.5935 |
| Mandya - Mysore | -485.0000 | 0.3386540923 | | -1584.0731 | 614.0731 |

**EXAMPLE 2**

**my.dat <-"C:/researchrdata/ agricolae_1.csv"**

**my.dat**

**attach (my.dat)**

```
> G
   ISOLATE YIELD
1  PRSV    2937
2  PRSV    3119
3  PRSV    2999
4  PRSV    3358
5  PRSV    3882
6  PRSV    2736
7  PLRV    2971
8  PLRV    3016
9  PLRV    3307
10 PLRV    4036
11 PLRV    3875
12 PLRV    3180
13 TMV     3238
14 TMV     2640
15 TMV     2834
16 TMV     3426
17 TMV     3676
18 TMV     3351
19 CMV     2725
20 CMV     4179
```

```
21  CMV     4196
22  CMV     4145
23  CMV     2697
24  CMV     3837
25  TOBSV   4046
26  TOBSV   2667
27  TOBSV   3176
28  TOBSV   4001
29  TOBSV   3978
30  TOBSV   2873
> my.model <-aov (Yield ~ Isolate, data = g)
> cv.model(my.model)
[1] 15.88272
> mean(Yield)
[1] 3370.033
> df<-df.residual(my.model)
> MSerror<-deviance(my.model)/df
> comparison <- LSD.test(Yield, Isolate , df, MSerror)
> LSD.test(model, "Isolate",console=TRUE)
> comparison <- LSD.test(Yield, Isolate , df, MSerror)
> LSD.test(my.model, "Isolate",console=TRUE)
Study: my.model ~ "Isolate"
LSD t Test for Yield
Mean Square Error:  286495.7
Isolate,  means and individual ( 95 %) CI
```

| | YIELD | STD R | LCL | UCL | MIN | MAX |
|---|---|---|---|---|---|---|
| CMV | 3629.833 | 723.7879 | 6 | 3179.791 | 4079.876 | 4196 |
| PLRV | 3397.500 | 451.3601 | 6 | 2947.458 | 3847.542 | 4036 |
| PRSV | 3171.833 | 404.1670 | 6 | 2721.791 | 3621.876 | 3882 |
| TMV | 3194.167 | 387.1043 | 6 | 2744.124 | 3644.209 | 3676 |
| TOBSV | 3456.833 | 625.8458 | 6 | 3006.791 | 3906.876 | 4046 |

alpha: 0.05 ; Df Error: 25

Critical Value of t: 2.059539

Least Significant Difference 636.456

Means with the same letter are not significantly different.

Groups, Treatments and means

| Groups, and means | Treatments | MEANS |
|---|---|---|
| a | CMV | 3630 |
| a | TOBSV | 3457 |
| a | PLRV | 3398 |
| a | TMV | 3194 |
| A | PRSV | 3172 |

> comparison <- LSD.test(Yield, Isolate, df, MSerror, group=F)

> outLSD <-LSD.test(my.model, “Isolate”, group=F,console=TRUE)

Study: my.model ~ “Isolate”

LSD t Test for Yield

Mean Square Error:  286495.7

Isolate,  means and individual ( 95 %) CI

| | YIELD | STD R | LCL | UCL | MIN | MAX |
|---|---|---|---|---|---|---|
| CMV | 3629.833 | 723.7879 | 6 | 3179.791 | 4079.876 | 4196 |
| PLRV | 3397.500 | 451.3601 | 6 | 2947.458 | 3847.542 | 4036 |
| PRSV | 3171.833 | 404.1670 | 6 | 2721.791 | 3621.876 | 3882 |
| TMV | 3194.167 | 387.1043 | 6 | 2744.124 | 3644.209 | 3676 |
| TOBSV | 3456.833 | 625.8458 | 6 | 3006.791 | 3906.876 | 4046 |

alpha: 0.05 ; Df Error: 25

Critical Value of t: 2.059539

Least Significant Difference 636.456

Means with the same letter are not significantly different.

Groups, Treatments and means

| Groups, and means | Treatments | Means |
|---|---|---|
| A | CMV | 3630 |
| A | TOBSV | 3457 |
| A | PLRV | 3398 |
| A | TMV | 3194 |
| A | PRSV | 3172 |

```
> comparison <- LSD.test(Yield, Isolate, df, MSerror, group=F)
> outLSD <-LSD.test(my.model, "Isolate", group=F,console=TRUE)
```

Study: my.model ~ "Isolate"

LSD t Test for Yield

Mean Square Error: 286495.7

Isolate, means and individual ( 95 %) CI

| | YIELD | STD R | LCL | UCL | MIN | MAX |
|---|---|---|---|---|---|---|
| CMV | 3629.833 | 723.7879 | 6 | 3179.791 | 4079.876 | 4196 |
| PLRV | 3397.500 | 451.3601 | 6 | 2947.458 | 3847.542 | 4036 |
| PRSV | 3171.833 | 404.1670 | 6 | 2721.791 | 3621.876 | 3882 |
| TMV | 3194.167 | 387.1043 | 6 | 2744.124 | 3644.209 | 3676 |
| TOBSV | 3456.833 | 625.8458 | 6 | 3006.791 | 3906.876 | 4046 |

ALPHA: 0.05 ; DF ERROR: 25

CRITICAL VALUE OF T: 2.059539

Comparison between treatments means

| | Difference | Pvalue sig | LCL | UCL |
|---|---|---|---|---|
| CMV - PLRV | 232.33333 | 0.4591802 | -404.1226 | 868.7893 |
| CMV - PRSV | 458.00000 | 0.1508195 | -178.4560 | 1094.4560 |
| CMV - TMV | 435.66667 | 0.1709203 | -200.7893 | 1072.1226 |
| CMV - TOBSV | 173.00000 | 0.5805848 | -463.4560 | 809.4560 |
| PLRV - PRSV | 225.66667 | 0.4720282 | -410.7893 | 862.1226 |
| PLRV - TMV | 203.33333 | 0.5165671 | -433.1226 | 839.7893 |
| PLRV - TOBSV | -59.33333 | 0.8492935 | -695.7893 | 577.1226 |
| PRSV - TMV | -22.33333 | 0.9429625 | -658.7893 | 614.1226 |
| PRSV - TOBSV | -285.00000 | 0.3652160 | -921.4560 | 351.4560 |
| TMV - TOBSV | -262.66667 | 0.4034056 | -899.1226 | 373.7893 |

```
> print(outLSD)
```

## $statistics

| Mean | CV | Mserror |
|---|---|---|
| 3370.033 | 15.88272 | 286495.7 |

## $parameters

| DF | NTR | T.VALUE |
|---|---|---|
| 25 | 5 | 2.059539 |

## $means

| | Yield | STD R | LCL | UCL | Min | Max |
|---|---|---|---|---|---|---|
| CMV | 3629.833 | 723.7879 | 6 | 3179.791 | 4079.876 | 4196 |
| PLRV | 3397.500 | 451.3601 | 6 | 2947.458 | 3847.542 | 4036 |
| PRSV | 3171.833 | 404.1670 | 6 | 2721.791 | 3621.876 | 3882 |
| TMV | 3194.167 | 387.1043 | 6 | 2744.124 | 3644.209 | 3676 |
| TOBSV | 3456.833 | 625.8458 | 6 | 3006.791 | 3906.876 | 4046 |

## $comparison

| | Difference | Pvalue Sig | LCL | UCL |
|---|---|---|---|---|
| CMV - PLRV | 232.33333 | 0.4591802 | -404.1226 | 868.7893 |
| CMV - PRSV | 458.00000 | 0.1508195 | -178.4560 | 1094.4560 |
| CMV - TMV | 435.66667 | 0.1709203 | -200.7893 | 1072.1226 |
| CMV - TOBSV | 173.00000 | 0.5805848 | -463.4560 | 809.4560 |
| PLRV - PRSV | 225.66667 | 0.4720282 | -410.7893 | 862.1226 |
| PLRV - TMV | 203.33333 | 0.5165671 | -433.1226 | 839.7893 |
| PLRV - TOBSV | -59.33333 | 0.8492935 | -695.7893 | 577.1226 |
| PRSV - TMV | -22.33333 | 0.9429625 | -658.7893 | 614.1226 |
| PRSV - TOBSV | -285.00000 | 0.3652160 | -921.4560 | 351.4560 |
| TMV - TOBSV | -262.66667 | 0.4034056 | -899.1226 | 373.7893 |

## 2 FACTOR rbd IN agricolae

> k <-read.csv(file= “C:/Users/dksamuel_now/Desktop/R-book/data/newanova.csv”, head=TRUE, sep= “,”)

> attach(k)

The following object is masked from g:

Yield

> k

| | Spray | Irrigation | Yield |
|---|---|---|---|
| 1. | Yes | Full | 10249 |
| 2. | Yes | Full | 12395 |
| 3. | Yes | Full | 14456 |
| 4. | Yes | Full | 11872 |
| 5. | Yes | Full | 11126 |
| 6. | Yes | Full | 10760 |
| 7. | Yes | Full | 11859 |
| 8. | Yes | Full | 13462 |
| 9. | Yes | Full | 14855 |
| 10. | Yes | Full | 11977 |
| 11. | Yes | Half | 10812 |
| 12. | Yes | Half | 11702 |
| 13. | Yes | Half | 13933 |
| 14. | Yes | Half | 13442 |
| 15. | Yes | Half | 11316 |
| 16. | Yes | Half | 11493 |
| 17. | Yes | Half | 14561 |
| 18. | Yes | Half | 11558 |
| 19. | Yes | Half | 12069 |
| 20. | Yes | Half | 12834 |
| 21. | No | Full | 11839 |
| 22. | No | Full | 14823 |
| 23. | No | Full | 10773 |
| 24. | No | Full | 10825 |
| 25. | No | Full | 10668 |

| | Spray | Irrigation | Yield |
|---|---|---|---|
| 26. | No | Full | 11139 |
| 27. | No | Full | 13325 |
| 28. | No | Full | 13965 |
| 29. | No | Full | 14646 |
| 30. | No | Full | 12605 |
| 31. | No | Half | 13828 |
| 32. | No | Half | 13756 |
| 33. | No | Half | 10374 |
| 34. | No | Half | 10720 |
| 35. | No | Half | 10982 |
| 36. | No | Half | 10197 |
| 37. | No | Half | 12552 |
| 38. | No | Half | 13985 |
| 39. | No | Half | 13285 |
| 40. | No | Half | 11787 |

```
> interaction.plot(Spray, Irrigation, Yield)
> k.model = lm(Yield ~ Spray + Irrigation + Spray*Irrigation)
> anova(k.model)
Analysis of Variance Table
Response: Yield
```

| | | | | | |
|---|---|---|---|---|---|
| Spray | 1 | 10791 | 10791 | 0.0049 | 0.9449 |
| Irrigation | 1 | 147987 | 147987 | 0.0665 | 0.7979 |
| Spray:Irrigation | 1 | 370755 | 370755 | 0.1667 | 0.6855 |
| Residuals | 36 | 80064467 | 2224013 | | |

```
> l.model = lm(Yield ~ Irrigation + Spray + Irrigation*Spray)
Error in eval(expr, envir, enclos) : object 'Sprtay' not found
> anova(l.model)
Error in anova(l.model) : object 'l.model' not found
> l.model = lm(Yield ~ Irrigation + Spray + Irrigation*Spray)
> anova(l.model)
```

Analysis of Variance Table

Response: Yield

| | | | | | |
|---|---|---|---|---|---|
| Irrigation | 1 | 147987 | 147987 | 0.0665 | 0.7979 |
| Spray | 1 | 10791 | 10791 | 0.0049 | 0.9449 |
| Irrigation:Spray | 1 | 370755 | 370755 | 0.1667 | 0.6855 |
| Residuals | 36 | 80064467 | 2224013 | | |

```
>> qqnorm(k.model$res)
> qqnorm(l.model$res)
> plot(k.model$fitted, k.model$res, xlab="Fitted", ylab="Residuals")
> plot(l.model$fitted, l.model$res, xlab="Fitted", ylab="Residuals")
```

Paired t-test

```
> REG = C(16, 20, 21, 22, 23, 22, 27, 25, 27, 28)
> PREM = C(19, 22, 24, 24, 25, 25, 26, 26, 28, 32)
> t.test(prem,reg,alternative="greater", paired=TRUE)
```

Paired t-test

data: prem and reg

t = 4.4721, df = 9, p-value = 0.0007749

alternative hypothesis: true difference in means is greater than 0

95 percent confidence interval:

1.180207 INF

sample estimates:

mean of the differences

2

```
> Control = c(91, 87, 99, 77, 88, 91)
> Treat = c(101, 110, 103, 93, 99, 104)
> t.test(Control, Treat, alternative= "less", var.equal=TRUE)
```

Two Sample t-test

data: Control and Treat

t = -3.4456, df = 10, p-value = 0.003136

alternative hypothesis: true difference in means is less than 0

95 percent confidence interval:

-Inf -6.082744

sample estimates:

mean of x mean of y

88.83333 101.66667

> x = c(0.593, 0.142, 0.329, 0.691, 0.231, 0.793, 0.519, 0.392, 0.418)

> t.test(x, alternative= "greater", mu=0.3)

One Sample t-test

data: x

t = 2.2051, df = 8, p-value = 0.02927

alternative hypothesis: true mean is greater than 0.3

95 percent confidence interval:

0.3245133 INF

sample estimates:

mean of x

0.4564444

## ANOVA SIMPLER WAY

Soluble_Boran A 94,89,81,82,88,95,96,98,93,88

Zinc_Sulphate B 84,79,74,72,71,75,82,83,88,79

Ferous_Sulphate C 64,66,65,63,62,70,69,68,69,61

Yield = c(94,89,81,82,88,95,96,98,93,88, 84,79,74,72,71,75,82,83,88,79, 64,66,65,63,62,70,69,68,69,61)

Spray = c(c(rep("Soluble_Boran" ,10), rep("Zinc_Sulphate ",10), rep("Ferous_Sulphate" ,10)))

Yld_data = data.frame(Yield,Spray)

Yld_data

plot(Yield ~ Spray, data = Yld_data)

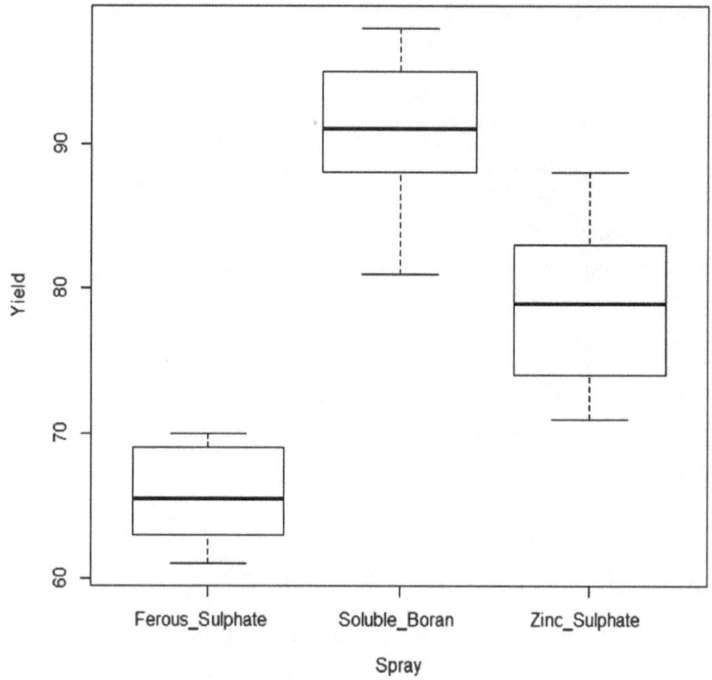


anova.anal = aov (Yield ~ Spray, data = Yld_data)

**summary(anova.anal)**

Df Sum Sq Mean Sq F value Pr(>F)

Spray 2 3053.3 1526.6 60.74 1.01e-10 ***

Residuals 27 678.6 25.1

pairwise.t.test(Yield, Spray, p.adjust="bonferroni")

Pairwise comparisons using t tests with pooled SD

data: Yield and Spray

| | Ferous_Sulphate | Soluble_Boran |
|---|---|---|
| Soluble_Boran | 5.1e-11 | - |
| Zinc_Sulphate | 1.1e-05 | 5.1e-05 |

anova.anal = aov (Yield ~ Spray, data = Yld_data)

TukeyHSD(anova.anal, conf.level = 0.95)

Tukey multiple comparisons of means

95% family-wise confidence level

Fit: aov(formula = Yield ~ Spray, data = Yld_data)

**$Spray**

| | DIFF | LWR | UPR | PADJ |
|---|---|---|---|---|
| Soluble_Boran -Ferous_Sulphate | 24.7 | 19.141089 | 30.258911 | 0.00e+00 |
| Zinc_Sulphate -Ferous_Sulphate | 13.0 | 7.441089 | 18.558911 | 1.05e-05 |
| Zinc_Sulphate -Soluble_Boran | -11.7 | -17.258911 | -6.141089 | 4.91e-05 |

**Factorial Data Analysis**

my.data <-read.csv("C:/Users/dksamuel_now/Desktop/R-book/data/tooth Growth_PGR.csv")

str(my.data)

'data.frame': 60 obs. of 3 variables:

$ len : num 4.2 11.5 7.3 5.8 6.4 10 11.2 11.2 5.2 7 ...

$ pgr : Factor w/ 2 levels "GA", "IAA": 1 1 1 1 1 1 1 1 1 1 ...

$ dose: Factor w/ 3 levels "low", "med", "high": 1 1 1 1 1 1 1 1 1 1 ...

my.data$dose = factor(my.data$dose,

levels=c(3,5,7),labels=c ("low", "med", "high"))

my.data[seq(1,60,5),]

replications(len ~ pgr * dose, data=my.data)

replications(len ~ pgr * dose, data=my.data [1:58,])

boxplot(len ~ pgr * dose, data=my.data, ylab= "Shoot Length", main= "Boxplots of Shoot Growth Data")

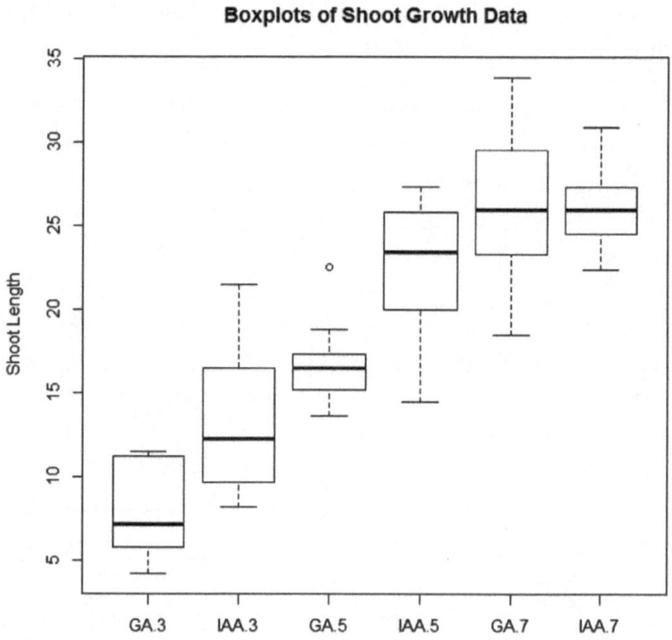


with(my.data, interaction.plot(x.factor=dose, trace.factor=pgr, response=len, fun=mean, type= "b", legend=T, ylab= "ShootLength", main= "Interaction Plot", pch=c(1,19)))

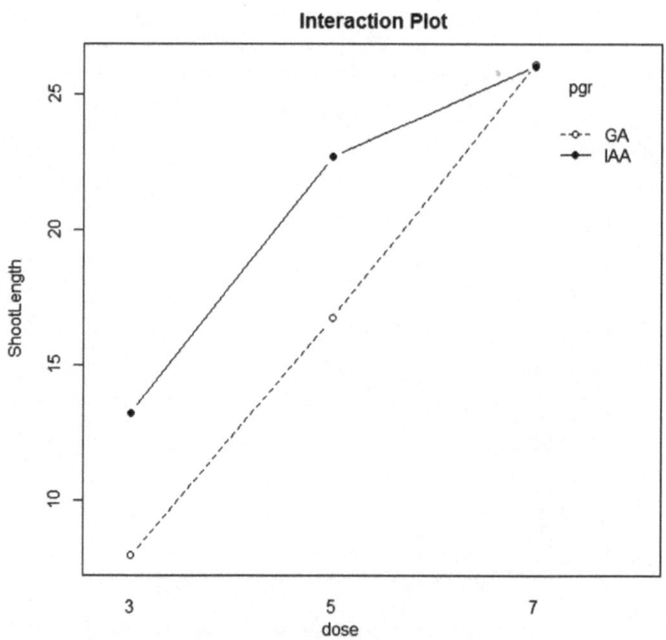

coplot(len ~ dose | pgr, data=my.data, panel=panel.smooth, xlab= "Shoot Growth data: length vs dose, given type of pgr")

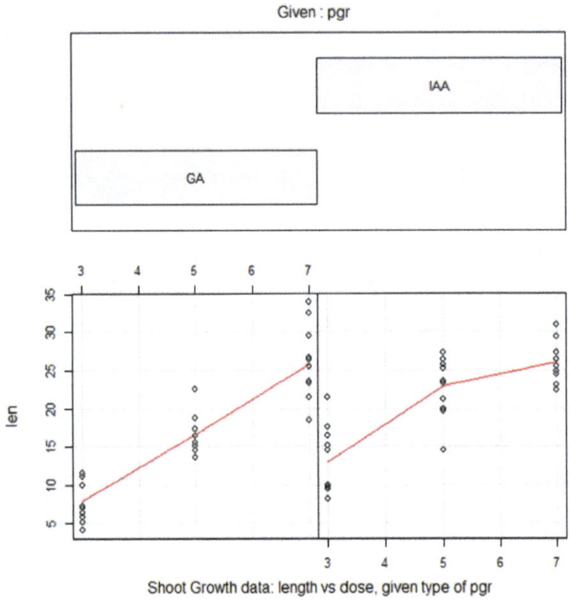


coplot(len ~ dose | pgr, data=my.data, panel=panel.smooth, show.given= FALSE, xlab= "Shoot Growth data: length vs dose, given type of pgr")

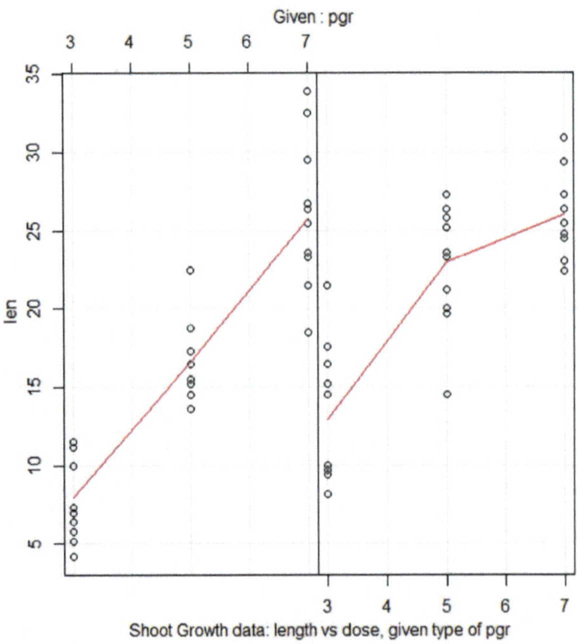

## 8.14 AUDPC

The area under the disease progress curve (AUDPC), calculates the absolute and relative progress of the disease. It is required to measure the disease in percentage terms during several dates, preferably equidistantly. AUDPC needs Agricolae

```
days<-c(7,14,21,28,35,42)

evaluation<-data.frame(E1=10,E2=40,E3=50,E4=70,E5=80,E6=90)

print(evaluation)

E1 E2 E3 E4 E5 E6

1 10 40 50 70 80 90

absolute <-audpc(evaluation,days)

relative <-audpc(evaluation,days,"relative")
```

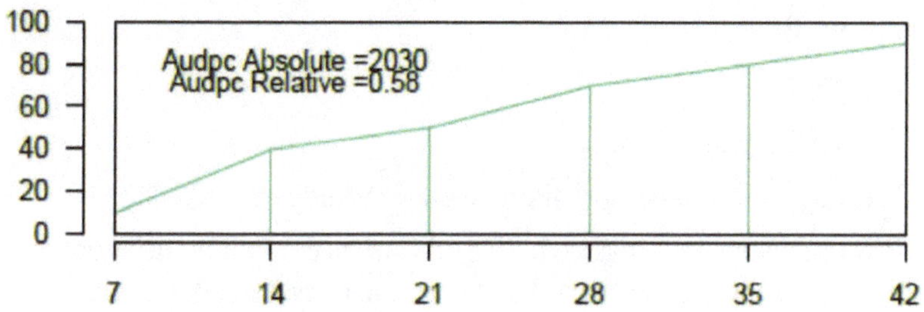


**Fig.** AUDPC: area under the curve

If you're using Excel for data analysis, give R a try. You'll be thankful you did.

*Further Reading*

http://www.michaelmilton.net/2010/01/26/when-to-use-excel-when-to-use-r/

http://www.burns-stat.com/first-step-towards-r-spreadsheets/

http://www.burns-stat.com/spreadsheet-r-vector/

http://www.burns-stat.com/documents/tutorials/spreadsheet-addiction/

http://blog.revolutionanalytics.com/2013/10/how-to-switch-from-spreadsheets-to-r-for-data-analysis.html

http://blog.revolutionanalytics.com/2013/04/more-reasons-not-to-use-excel-for-modeling.html

http://blog.revolutionanalytics.com/2013/02/did-an-excel-error-bring-down-the-london-whale.html

http://robjhyndman.com/hyndsight/rvsexcel/

http://andrewgelman.com/2013/04/17/data-problems-coding-errors-what-can-be-done/

http://andrewgelman.com/2013/04/17/excel-bashing/

http://r-dir.com/blog/2013/11/r-vs-excel-for-data-analysis.html

http://www.quantumforest.com/2013/12/excel-fanaticism-and-r/

http://christophergandrud.blogspot.com/2013/04/reinhart-rogoff-everyone-makes-coding.html

http://r4stats.com/articles/popularity/

http://www.nytimes.com/2009/01/07/technology/business-computing/07program.html?pagewanted=all

**Features of R**

- R is free.
- R is well documented.
- R runs (really well) on *nix as well as Windows and Mac OS.
- R is open source. Trust in the R software is evident by its support among distinguished statisticians. However, the R user need not rely on trust, as the source code for R is freely available for public scrutiny.
- R has a much broader range of statistical packages for doing specialist work.
- R has an enthusiastic user base who can offer helpful advice for free.
- R creates far better graphics than Excel.
- R has certain data structures such as data frames that can make analysis more straightforward than in Excel.
- R is better for doing complex jobs.
- R is a better educational tool as it uses standard statistical vocabulary rather than home baked terminology.
- R is easier to learn, use, and script than Excel.
- R allows students easily to work with scripts, thus allowing the work to be reproducible.
- R is intended to lead students towards programming; Excel is designed to keep people away from programming and encourages them to rely on

someone else doing their programming (and often their thinking) for them.

- The statistical package available in Excel is very limited in capability and should only be used by experienced applied statisticians who can work out when its output should be ignored.
- While R takes a while to learn, it provides a broad range of possible analyses and does not constrain users to a very limited set of methods (as is the case for Excel).

FSC
www.fsc.org
MIX
Papier | Fördert
gute Waldnutzung
FSC® C083411

Zeitfracht Medien GmbH
Ferdinand-Jühlke-Straße 7
99095 Erfurt, Deutschland
produktsicherheit@kolibri360.de